THE BERRY BOOK

Anne-Marie Driediger

hancock
house

ISBN 0-88839-267-2
Copyright © 1995 Anne-Marie Driediger

Cataloging in Publication Data
Driediger, Anne-Marie.
The berry book

ISBN 0-88839-267-2

1. Cookery (Berries) 2. Berries. I. Title.
TX813.B4D74 1995   641.6'47   C95-910512-3

Printed in Canada–Hignell

Production: Suzanne M. Chin

Published simultaneously in Canada and the United States by

HANCOCK HOUSE PUBLISHERS LTD.
19313 Zero Avenue, Surrey, B.C. V4P 1M7
(604) 538-1114   Fax (604) 538-2262

HANCOCK HOUSE PUBLISHERS
1431 Harrison Avenue, Blaine, WA 98230-5005
(604) 538-1114   Fax (604) 538-2262

TABLE OF CONTENTS

All For The 'Love Of The Land' 5

Pesky Critters or IPM? 14

All About Berries 17

Making Jam and Jelly 25

Where Did I Go Wrong? 32

Remaking Procedures 34

Preserving Berries 36

Appetizers—*Dressings, Dips, Salads and Soups* 43

Main Courses—*Main dishes, Salsa, Sauces, Relish* 52

Desserts—*Cakes, Cheesecakes, Shortcakes, Muffins* 59

   Crisps, Crumbles, Cobblers, Buckles 71

   Puddings and Parfait 79

   Sauces and Toppings 87

   Pies and Tarts 90

   Clafoutis, Flans and Tortes 106

   Desserts; layered, chilled, etc. 110

Preserving—*Jams, Jellies, Conserves, Vinegars* 122

Miscellaneous—*Bread, Breakfast and Drinks* 137

Charts and Conversions 145

Acknowledgments 154

Index 155

1ˢᵀ GENERATION BROTHERS

GEORGE

PETE

2ⁿᵈ GENERATION BROTHERS

DRIEDIGER FARMS

5

MURRAY ~1994~ MICHAEL

JUNE, THE PRES'

ANNE-MARIE, MURRAY'S WIFE

ALL FOR
THE 'LOVE OF THE LAND'

June Driediger

Born in June 1933 to a Saskatchewan wheat farming family and having wonderful memories of our family farm, is it any wonder that I fell in love with 'this land' when I came to British Columbia at Christmas time 1952? I had always loved eating strawberries, but never imagined I would spend the next forty plus years living and raising children on what was to become one of the largest straw-berry farms in B.C.

The summer of 1954, my husband, George and I came back to B.C. from the prairies, to help Dad and Mom Driediger harvest their strawberry crop. Dietrich and Paulina Driediger had come with their family to B.C. from the prairies in 1944. They started growing strawberries in the 'Otter District' on a 10-acre farm on 56th Avenue. At that time, 6 to 8 acres were in berries. We came home to help that summer and stayed on to help during the fall months. That was the beginning of our farming years! We took over the farm the following spring and from then on, strawberry farming in the Otter District began to change.

Between the years of 1954 and 1957, our farm grew from 8 acres to 25. Then by 1963, we were farming approximately 250 still within a few miles of the original home place. Some of this land was leased and some had been purchased.

George's brother Pete Driediger, and his wife Hilda, joined us in the early 1960s. In April of 1964, we formed a company called Driediger Bros. Farms Ltd.

5

During those early years of farming everything was done by hand. In early spring we would dig and clean the young plants for planting. First the fields were worked with a hand rototiller, then we would take twine and roll it out from one end of the field to the other. This was a must, in order to have perfectly straight rows. Then planting began with a spade and buckets of young plants. Can you imagine how many back-bends one had to do to plant 6,000 plants per acre! After the planting was finished, we fertilized, hoed, cut runners and blossoms—all by hand. We harvested the following year.

During these years of expansion, we were fortunate to have several hard working people join us, including the Suderman family, Elaine "Peppy" Peplow, Irene Warkentin, George "Big George" Wiebe, Ann "Big Annie" Krochmoly, Mike and Olga Dudzak and Mary Jones. Their love of and dedication to the farm was a great part of our success. Most of these people worked for us for over 20 years, and Big George is still with us today.

The late 1950s through to the 1970s saw many changes…from hand planting to machine planters, updated tractors and forklifts. In the early 1950s, the British Sovereign berry was grown by all the local farms. It was tasty but very seedy and hard to pick. We were one of the first farms to try a new variety call the Siletz—a bright red, sweet berry and very easy to pick. We grew the Siletz for several years along with other new varieties like the Northwest and Peugot Beauty. Today's main varieties are Totem and Raineer.

With so many acres under cultivation, we now needed hundreds of pickers. Soon, used school buses were purchased to pick up school children and adults from Coquitlam, New Westminster, Cloverdale and Langley to pick on our farm. School would be let out early to accommodate the season. This system meant paying each picker in cash every night before they boarded the bus for home. There were no guarantees they would return the next day. For a few years, we had a canteen on the fields selling hot dogs, pop and ice cream— what a circus some days!

During the 1960s, Hilda and I started the "U-Pick" that our farm has become so famous for. It all started out in the fields where the wooden crates filled with strawberries were stacked. People came from miles around with their containers and bought the fruit right off the field. Soon we were letting them pick in the fields after the fruit was no longer economical to harvest. Today, the U-Pick fields are grown specifically for the public with no commercial picking in them.

During the early years of farming, George and I had four children; Murray, Brenda, Rhonda and Michael. As small children, they spent many hours in the fields playing in the rows. As they grew older, they learned early the value of hard work, dedication and developed a sense of pride which has served them well in their lives and careers today.

In 1968, we bought the shares of Driediger Bros. Farm from Pete and Hilda, and continued to farm until 1980. In 1981, George and I parted company. I purchased the farm from him and formed a new partnership with my two sons.

Since this new partnership was formed, we have taken this company in new directions. We have added several commodities to our product line with continued harvest from June to December. We have constructed 'state of the art' cooling facilities. This, coupled with improved packaging and the availability of Air Cargo, has enabled us to ship our products Canada-wide.

It is with great satisfaction and complete confidence I am now passing on to them and to their generation, the original "Driediger Bros. Farm."

Yes, it was *'for the love of the land'* and *'for the love of a family heritage.'*

OUR FIRST FRUIT STAND ~ 1976

TODAY'S FRUIT STAND WITH BUILT-IN COOLING FACILITY

SHIPPING TO THE JAM PLANT ~1971

EARLY CUSTOMERS USED TO WALK RIGHT INTO THE FIELDS TO BUY ~1966

OUR FIRST NEW TRUCK ~1957

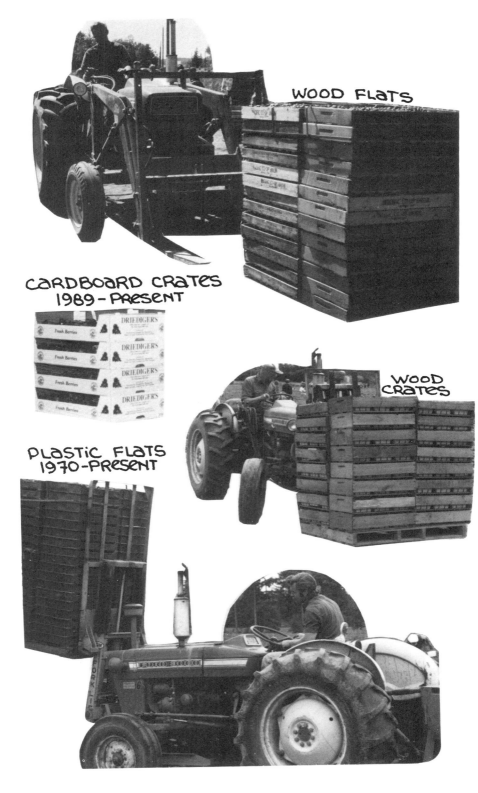

WOOD FLATS

CARDBOARD CRATES
1989 - PRESENT

WOOD CRATES

PLASTIC FLATS
1970 - PRESENT

MIKE & OLGA DUDZAK

ANNIE KROTCHMOLEY

IRENE WARRENTINE

GEORGE WIEBE

RANDY McEWEN

JASBIR JOHAL

MARY JONES

ELAINE (PEPPY) PEPLOW

MACHINE HARVESTING GOOSEBERRIES

STRAWBERRIES

CAULIFLOWER

U-PICK

CORN AND

BLACK

CURRANTS

BLUEBERRIES

1994
BLAST COOLING FACILITIES
FOR WHOLESALE
SHIPPING
ALL ACROSS CANADA!

PESKY CRITTERS or IPM?

Dr. Debbie Henderson and A. M. Driediger

What is Integrated Pest Management?

Integrated Pest Management (IPM) is a farming practice that minimizes the use of pesticides for control of disease and pest problems in crops. Various methods are used to reduce damage to the crop by pests and disease. *The use of pesticides is employed only when absolutely necessary.*

A Little History First...

Prior to 1945, few pesticides were available and most pest and disease management for food crops consisted of:

* Choosing varieties resistant to local area problems.
* Cultural practices such as rotation, "intercropping."
* Mechanical methods like cultivating between the rows of a crop to control weeds or flooding cranberries to kill pests.
* Biological controls such as introducing 'good' bugs that kept the pest bugs at low levels.
* Natural source pesticides like copper, sulphur, lead arsenate and pyrethrum.

The result, yields were low, quality was poor and there were predictable losses each year due to disease and pest problems. Varieties resistant to certain diseases or insect damage were not noted for their high yields. The family farm fed the family with only a little left over.

In 1946, the world changed. DDT was introduced to U.S. agriculture. It was a magic bullet, so devastating to insect populations that

14

complete eradication of pest problems seemed possible. So effective was it against mosquitoes, that the World Health Organization planned to eradicate malaria from the globe using DDT. Its discoverer received a Nobel Prize.

DDT became a standard remedy for any pest problem going. With its many early successes, farmers became addicted to the quick fix of the pesticide approach. Crop yields soared dramatically and quality of crops greatly improved. Old plant varieties that were once favored because of their disease or pest resistance, were passed over in favor of high yielding varieties that depended on high inputs of chemical pesticides and fertilizers. Many of the cultural methods used were no longer necessary. This magic however was not to last.

The rude awakening came not long after the introduction of pesticides. It was found that using pesticides against one pest also destroyed the natural control agents of other insects which were not previously considered pests. Now more sprays were required to control the new pest.

To make matters worse, pests and diseases began showing resistance from frequent exposure to pesticides. This led to more and different chemicals produced, higher concentrations used, more frequent sprays applied or mixes of chemicals used to control the same pests. The pesticide treadmill began as the dramatic yields experienced in the early pesticide era could not be maintained without the use of more pesticide applications.

Can We Change?

Economics drives decisions made at the farm level. Farmers take all the risks for weather, are responsible for the care of the soil and must absorb the increased costs of production. They are subject to the whims of the market as with any other business. Historically, the price a farmer receives for his crop remains constant over many years while inflation and cost of labour and production continue to rise. Therefore, economics and effectiveness will determine the use of pesticides versus alternative cultural practices. In the last few years it has finally been possible to save a farmer production costs, by offering an alternative.

A New Age Alternative....

This alternative is known as Integrated Pest Management. We at Driediger Farms began implementing IPM as early as 1985. We are committed to reducing our use of pesticides and adopt as many alternative strategies as possible for controlling diseases and pests for each crop. The key is to integrate in an organized manner using all available techniques for keeping pests and diseases below economically damaging levels, while minimizing environmental impact and remaining cost effective.

How Can We Implement IPM?

One of the most important tools of the IPM approach is to monitor the crops regularly to determine *if* there is a pest or disease problem, and *where* precisely. In the past, any insect pest found in a field was considered a problem which had to be eradicated. Now, in an IPM field, it is recognized that pest species can be present in numbers that may not damage the crop. Therefore, pests are treated *only when and where necessary.*

We at Driediger Farms are concerned about the unnecessary use of pesticides. Through IPM, the use of plastic and sawdust mulches help reduce weeds in place of using herbicides. Predator insects and biocides have also been introduced. These methods have reduced the amount of synthetic pesticides dramatically while maintaining both quality and economically acceptable yields. We have been an enthusiastic participant on any research on alternative methods for pest control and will continue to be, in the future. At any given time, the farm has 3 or 4 different research projects in progress.

As further breakthroughs are made, we will be implementing them into our operation with the aim of eliminating chemical pesticides...WE CARE.

ALL ABOUT BERRIES

Anne-Marie Driediger

A berry is considered a soft fruit. Because it is, berries have the shortest shelf life of all the fruits. *Berries are very perishable.* The ideal situation would be to pick your own and pop them into your mouth for ultimate freshness. Next in order of desirability:

Pick -Your-Own: There are many U-Pick farms available in the Lower Mainland. The best places offer clean, supervised fields thus ensuring the best picking possible in the least time. Supervised fields offer the picker a row(s) plentiful with fruit with no one else in it! Often children are welcome but shouldn't be allowed to roam free, possibly damaging the plants. Bringing your own containers helps reduce the cost per pound. Picking your own allows you to be the quality controller. Remember, the smaller strawberries are usually the sweetest. Picking your own fruit is often considered a privilege by the farmers so please take care of those delicate plants and pick your row clean. *All ripe fruit is best regardless of size.*

Putting too much fruit in one container will result in squashed fruit on the bottom. Plastic ice cream pails are often the most handy since they all weigh the same (¼ lb.-½ lb.), holding 5-6 lbs. of fruit (not too deep), and possessing a handle. Pick with the stems on for strawberries and keep the handling down to a minimum. Raspberries are best packed in a more shallow pan than an ice cream pail since there is no core to help keep their shape (3-4 berries deep only). Wash fruit before eating.

So take your time, enjoy the scenery, the weather, and remember, the back pain does go away! Be prepared for all kinds of weather; boots, sun hats, gloves and a wet cloth to clean up with. Bring a picnic lunch, the family, and make an outing of it!

Buying Fresh: Farm fresh is always the best. Many farms in the Lower Mainland offer fresh picked berries in prepackaged quantities. If you just want a basket or bulk for processing, choose a farm that provides precooled berries. On Driediger Farms we whisk the berries off the field as quickly as possible and store them in our precooled facility. This process extends the shelf life. Sorted berries are then packaged on the premises to guarantee freshness and quality.

Plan in advance to make the fruit purchase your last stop before home. Travelling time in non-refrigerated vehicles for long periods reduces the shelf life of your berries. Once at home, tend to your berries immediately. If eating fresh, store them in the refrigerator unwashed and wash as you use them. If processing or freezing your berries, do so that day, preferably immediately.

Driediger Farms is devoted to freshness and quality but if you want shelf life, look after them today! Apples and oranges can be stored in a cool, dry place for weeks, but not berries. If you plan to eat, can, jam or freeze fruit this summer, do so in order of their season. Buying several types at once in large quantities is not recommended. Make several trips and process as you go. Ensure you are getting each fruit in its peak season. Have all the equipment and ingredients ready before you purchase the fruit.

Our Approx. Seasons and Varieties:
Strawberries: Jun - mid-July (Totems/Raineer)
Raspberries: July - mid-Aug (Skeena/Meeker/Chilcotin/Comox)
Blueberries: July - mid-Aug (Early-Blue/Duke/Blue-Crop/Hardy-Blue/Jersey)
Black Currant: July-mid-Aug (Ben-Loman/Ben-Nevis)
Red Currant: July-mid-Aug (Jonkheer Van Tetes/Red Lake)
Gooseberries: July - mid-Aug (Hinnonmaeki Red & Yellow)
Blackberries: July - mid-Aug (Lochness/Kotata/Chester)

STRAWBERRIES

Red, juicy and pulpy, strawberries belong to the genus *Fragaria*, a member of the rose family. This fruit varies in size, color, shape and taste according to the variety. Great strides have been made in the research of varieties at the University of British Columbia and the Aggasiz Research Station. Some varieties are grown specifically for fresh market and others for processing. Here in the Lower Mainland, the berries you will find are vine-ripened, red and juicy throughout the entire berry. Taste the difference!

Strawberries have been cultivated to a limited extent in European gardens since the early 14th century. Most Europeans ate the wild strawberry, known as the Alpine strawberry (Fragaria vesca), which was a native of northern Italy. It is this fruit that was mentioned by the Roman writers Virgil, Ovid and Pliny. The love affair with strawberries continues today and the month of June is a magical time with the end of school, the beginning of summer and fresh strawberries.

Eaten plain, dipped in sugar or with chocolate, strawberries are a treat. Made into preserves, they continue to be one of the most popular jams today. Strawberries enjoy the largest consumption of all berries worldwide. The strawberry shortcake is a beloved classic of the strawberry season inspiring many festive occasions.

Storage: Pick or buy fresh, sort berries and refrigerate immediately. Do not wash until ready to use. Those not used for fresh eating should be processed that day.

Washing: Keep stems on and lightly rinse in colander. Drain well and place on paper towels to dry. Stem using stemming tool, fingers, knife or straw. Slice, quarter or leave whole with or without sugar.

RASPBERRIES

Grown on a bush, the raspberry is a fruit of the *Rubus* genus which is a member of the rose family. The raspberry is made up of many small drupelets in contrast to all the other berries which retain their stems or receptacles when the fruit is picked. The stem of the raspberry separates from the berry and remains on the plant when it is ripe. It is for this reason that the raspberry is the most delicate of all the berries and should be handled with great care.

Raspberries may be red, purple, black or amber in colour. They are very juicy and can be slightly tarter in taste than strawberries. This fruit is interchangeable in many recipes calling for other soft fruit. It can be eaten fresh or frozen for later. The berries make great preserves, jams, jellies, puddings and pies. When thinking dessert, think raspberries.

The Lower Mainland is the raspberry capital of Canada. Lower Mainland growers are some of the leading exporters of fresh and processed raspberries in the world. Raspberry juice is a popular commodity and many frozen products are produced from raspberries. The U-Pick is popular with raspberries, as some people feel it is easier to pick, but bring along a little person to help pick those lower on the bush!

Storage: Pick or buy fresh. Sort and refrigerate immediately. Keep them in a shallow container. Do not wash them until ready to eat or process. Those berries left for processing or freezing must be done the same day as purchased.

Washing: Raspberries are grown on bushes, unlike strawberries that lie on the ground. Therefore, after careful sorting, removing all leaves, some people prefer not to wash their raspberries because they are so delicate, especially if they are freezing them whole. However, it is recommended that all berries be rinsed lightly in a colander. Drain well and leave to dry on paper or old tea towels.

BLUEBERRIES

Ranging in colour from purplish-blue to blue-black, blueberries belong to the *Vaccinium* genus. There are many varieties of blueberries. They grow singly or in clusters on bushes from one to twenty feet high.

In their wild state, blueberries are found from the northern tip of Alaska down to Florida. They also grow profusely on other continents, even above the Arctic Circle. They are much prized wherever they grow. Today blueberries remain a popular fruit.

Blueberries are First Nations food. They were a major food supply for many tribes who ate them fresh or cooked with meat, or dried them in large amounts for winter.

Blueberries are the only pleasant aftermath of a forest fire, because they will grow in abundance on burnt-over ground. Today, cultivated blueberries are a real treat to eat fresh and are compatible in most recipes. Blueberries make excellent jam, pies, jelly, muffins, and many desserts. They freeze well too. It is not necessary to Individual Quick Freeze blueberries since they easily break away from the clump when frozen.

Storage: Pick or buy fresh in small or bulk quantities. Do not wash until ready to eat. Sort and remove any stems or leaves. Refrigerate until ready to eat. Remember, if you are processing berries, do it the same day for best results. Keep handling of blueberries down to a minimum to increase shelf life.

Washing: Rinse lightly in colander, drain and let dry on paper towels or old tea towels. For freezing individually, some people don't wash them since they grow on bushes. We always recommend that you wash your berries.

CURRANTS, BLACK & RED

The currant name is applied to two different fruits. One, a fresh currant, is a berry of the genus *Ribes* and a member of the gooseberry family. The other, a dry currant, is a dried grape of the genus *Vitis*. The fresh currant is first mentioned as a garden fruit in the 15th century. It resembles and probably took its name from the dry currant, which had been cultivated for centuries before that time.

These tiny, sweet-tart berries are native to the colder regions of both Europe and America and come in red, white and black varieties. Usually not available for U-Pick, red and black currants can be purchased directly on the farm.

Black currants are probably renowned for jelly and juice but are also nice for jams. They possess the highest vitamin count of all the berries and no doubt have been used for medicinal purposes throughout the ages.

Red currants are eaten fresh as a fruit as well as cooked in jams and jellies. Red currants are by far the best known and most frequently used. They make excellent jelly and are great for pies and desserts.

White currants are used in salads and fruit cups.

Storage: Buy fresh, refrigerate immediately. Process all currants that are not to be eaten fresh. Currants may be washed before storing them if they have been thoroughly dried. Eat soon.

Washing: Lightly rinse fruit in colander. Drain and remove strigs (stems) either by hand or with the aid of a fork. To remove strigs using a fork, hold strigs with one hand and place the fork into the top of the clump and gently pull down. Lightly rinse again to remove any bits of leaves or strigs. Drain and let dry on old tea towels or paper towels.

GOOSEBERRIES

This round, juicy berry is a first cousin of the fresh currant and belongs to the *Ribes* family. The bushes they grow on can be either thorny or thornless depending on the variety. Their flavour is tart, yet sweet when fully ripened and varies in size. There are a number of varieties; red, white, green, yellow and hairy.

Gooseberries are native to Europe and Asia and have been cultivated since the beginning of the 17th century. They are widely used in the northern countries of Europe. In English home cooking, gooseberries are prized in tarts, jams, canned, pies or as a sauce. The varieties used for cooking are the smaller, tarter ones.

The great yellow and red gooseberries are often used for fresh eating when fully ripened. Many remember the symbolism of gooseberries, bringing to mind a perfect summer day, filled with buzzing bees, scents of flowers and fruit mingling in the still air.

Storage: Buy fresh, clean, dry and store in a refrigerator or leave out at room temperature. Eat soon. Process today.

Washing: Remove all the flower ends and strig-bits. This is best done dry. Carefully rinse berries in a colander making sure to remove all the little bits! Drain well and leave to dry on old tea towels or paper towels.

BLACKBERRIES

This conical fruit is composed of many small fruits called drupe-lets. Blackberries are also called brambles, since they are the fruit of various brambles of the *Rubus* genus.

Like the raspberry, the stem or receptacle holding the blackberry remains part of the shrub. Ripe blackberries are purple-black in colour. Unripened or green blackberries are red. They may be wild or cultivated.

Blackberries are one of the most abundant wild crops in America. They have long been popular in England where every farmer's wife went black-berrying in September to turn her crop into puddings, pies, jams, jellies and even homemade wine.

Blackberry wine and cordial used to be a precious homemade drink. Their juice tastes wonderful when mixed in equal quantities with lemonade.

Storage: Pick or buy fresh, sort and refrigerate immediately. Keep them in a shallow container. Do not wash them until ready to eat or process. Those berries left for processing or freezing should be done up the same day.

Washing: Remove all leaves and rinse lightly in small quantities in a colander. Drain well and leave to dry on paper towels or old tea towels.

"If you have a habit of being attentive and expressing interest, your children will not confuse your loving instruction with rejection"

MAKING JAM & JELLY

MAKING JAM

There is great satisfaction in bringing out a jar of homemade jam! The taste will bring back the memories of summer. Apart from making a thoughtful gift, the jam quality is better than anything you can buy and you can save money too. Buy your fruit in season and enjoy it all winter. Making jam is really quite easy if you follow the instructions. If this is your first time, read the text first, select the recipe you want and follow it carefully. Make it with a friend, it can be a lot of fun!

Sweet, spreadable fruit preserves take many forms. Here is a description of the most common names:

Conserve: Small pieces or whole small fruit uniformly distributed in a thick sauce. May include nuts or raisins.
Jam: Crushed fruit cooked with or without sugar and pectin.
Jelly: Fruit juice cooked with or without sugar and pectin.
Marmalade: Citrus peel cooked with fruit juice and sugar.
Preserves: Small or whole pieces of fruit uniformly distributed in a thick sauce.
Spreadable Fruit: Soft, jam-like consistency with reduced sugar content.

Make one recipe at a time—DO NOT DOUBLE!!

EQUIPMENT: Use a large saucepan of stainless steel or enamel— aluminum is not recommended. Never fill more than half the saucepan since boiling jam bubbles and spits near the setting point. Use only a wooden spoon to stir the jam—plastic or metal is not recommended.

Cooked jam is best stored in glass jars. The paraffin method of sealing is outdated and not recommended. Use the metal lid and screw cap method using mason jars only. If you are making freezer (no-cook) glass jars are preferable to plastic containers. If you plan to give jam away as gifts, think about using some fancy jars to dress it up. The size of jar depends on how fast the jam will be consumed later. Smaller jars may take more time initially, but they make better gifts and once the jam is opened, there is less chance of spoilage if consumed sooner. Quart jars are not recommended.

FRUIT: Always buy fresh at the source. The fruit should be ripe to slightly underripe, as this is when pectin is at its highest. *Pectin* is a natural gum-like substance in some fruits; when boiled with it forms into a jelly. *Acid* helps extract the pectin, brighten the colour, improves the flavor and helps prevent crystallization. All fruits vary in the amount of pectin and acid:

Fruits High in Pectin & Acid: Black & Red Currants,
Gooseberries
Fruits Low in Pectin & Acid: Blackberries, Raspberries,
Strawberries

Pectin and acid can be easily added to low-pectin fruits in the form of citrus juice. You can mix a fruit high in pectin, like apple with one low in pectin. Therefore, there are two ways to make jam or jelly; *Naturally (no artificial pectin) or with artificial pectin.*

Test for Acid: Mix 1 tsp. lemon juice, 3 tbsp. water and ½ tsp. sugar. Taste this mixture and compare it to the taste of your fruit juice. If juice is sweeter, adjust taste by adding 1 tbsp. lemon juice to each cup fruit juice.

Test for Pectin: Gently shake 1 tsp. juice with 1 tbsp. rubbing alcohol in a closed container. *DO NOT TASTE.* Adequate pectin is present when mixture forms a jelly-like mass that can be picked up with a fork. If this does not occur, there is not enough pectin to form a gel. In such cases, select a recipe with artificial pectin.

The fruit needs to become soft before the sugar is added. The process of softening breaks down the cell walls of the fruit and releases the pectin. Generally, the fruit is brought gently to a boil and then allowed to simmer from 30 to 60 minutes to soften the fruit (for natural method). Sometimes extra water is added to prevent burning; the amount needed depends on the water content of the fruit and the quantity in the saucepan.

SUGAR: Sugar is very important because it preserves the fruit by retaining the natural fruit flavour and colour. It also enables it to set; too little will prevent the jam from setting, too much will darken and sweeten the jam. Use granulated, preserving or superfine sugar, as unrefined and raw sugars will smother the flavour of the fruit. Light corn syrup or honey will do this too.

Some recipes for natural jams will require the sugar to be warmed. Warming enables the sugar to dissolve faster and is done when using fruits that need to be boiled for only a short time. To do this, put it in a baking dish, spread it out and put the dish in a slow oven for 10 minutes.

The golden rule for jam making, is slow and long cooking to soften the fruit before adding sugar. Then very fast and short cooking as soon as the sugar has dissolved.

THE SETTING POINT: The setting point is the exact time to finish the cooking. The jam will not set properly if it does not reach this point. If the cooking goes beyond this point, the jam will darken and crystallize. There are three ways to determine the setting point:

Saucer Method: Take a small saucer cold from the freezer and drop some jam onto it. As the jam cools, it should set and crinkle if you push it around. Turn the plate upside-down. If the jam still sticks, the setting point is reached.

Spoon Method: This test is for jelly only. Dip a *metal* spoon into the jam. Remove and hold the spoon horizontally until the jam is slightly cooled. Turn the spoon gently. If the jam falls off in heavy flakes, it is at the setting point.

Temperature Method: Use a sugar thermometer. When the jam reaches 221°F, the setting point is reached.

BOTTLING: As soon as the setting point is reached, remove the saucepan from the heat and *remove any scum* that may have formed. *Allow to stand* for 10 to 15 minutes so the fruit distributes evenly through the jam. *Pour the jam* into clean, sterilized jars. *Sterilize* by using the 'open kettle method': put clean jars in a canner, cover jars with water and bring to a boil. Boil for 15 minutes.

Leave a head space between the top of jam and top of jar: ¼ inch for half-pint glass jars and ½ inch for pint jars. *Remove any air bubbles* in the jam by inserting a spatula along the sides. Air bubbles harbor bacteria and can cause discoloration of the surrounding jellied product. It can also interfere with obtaining an airtight seal.

SEALING: Wipe the jars down, wipe on and inside the rim. Seal immediately:

Paraffin wax: "Home canning has changed over the years. Paraffin wax is now considered unsafe."—USDA

2-Piece Metal Lids: Boil the lids in water on the stove. Place a boiled metal lid on top of jar and quickly screw band in place just until fingertip tight. Boiling lids soften the sealing compound. *Always use a new boiled lid,* it is only good for one time, but the screw bands can be reused for years. Process in water bath 10 minutes.

Water Bath Method

1) Set water bath canner with rack on stove. Fill with 4-5 inches of water. Cover and start heating over high heat. Also start heating additional water in kettle to fill canner after jars are in place.

2) Prepare sugar syrup if using, keep warm but not boiling until ready to use.

3) Prepare fruit: sort, wash and stem.

4) Fill the jars and place each one in canner. Make sure jars do not touch. Replace cover on canner each time you add a jar.

5) When the last jar has been added, fill canner with boiling water so water is 1-2 inches over top of jars. Cover canner. Heat water to a brisk rolling boil.

6) Start counting the processing time: Pints-15 min., Quarts-20 min. Adjust heat under canner so water boils gently during entire processing time. Add boiling water if water level drops. If boiling stops when you add water, stop counting processing time, turn up heat and wait for a full boil before resuming counting.

7) When time is up, turn off heat. Carefully remove hot jars with a jar lifter or long-handled tongs. Transfer hot jars to rack to cool out of draft. Allow air to circulate around jars. Do not cover or turn jars upside down.

8) *Do not move jars for 12 hours.* Wash cold and sealed jars. Wash and dry bands if removing.

9) Date and Label. Store.

STORING: After jars have been processed for sealing, *do not move jars for 12 hours.* Moving can break the gel, especially jellies. After 12 hours cooling, check the seal (sealed lids curve downward). Remove the screw bands if desired, *label and date.* Store in a cool (not subject to freezing), dry, dark place.

SHELF LIFE: For all your jam, jelly and canned fruit, 1 year's shelf life is about it. The flavour and quality of berries begins to decrease within a few months. Once opened, keep refrigerated and consume within 2 weeks for canned fruit and up to 4 weeks for jams and jellies.

GIFTS: Any jams, jellies, preserves etc. are considered *gifts of the heart.* For Christmas, to cheer up a friend, a thank you to someone special, or a gift for a teacher, giving something from your kitchen is always appreciated. You can cover the jars with paper or material and tie it with a ribbon. Remember, "it's the thought that counts." "It's the time that is appreciated!"

MAKING JELLY

A good fruit jelly is bright and clear and set but still a little wobbly. The fruit taste should be noticeable. Jam uses crushed fruit but jellies use the strained juice from the cooked fruit that is boiled with sugar to setting point. It is advisable to read 'Making Jam' before embarking on jellies.

Make one recipe at a time—DO NOT DOUBLE!!

FRUIT: The most suitable fruits are currants and gooseberries because they are high in pectin and acid. The other berry fruits generally need to be mixed with apple if you are making natural jelly (no artificial pectin). The fruit should be fresh and just underripe. As the yield is much less in jellies, buying the fruit in bulk or at reduced prices make it more economical.

Wash the fruit carefully, but don't worry about strigs or stems since the pulp is going to be strained. The fruit is cooked in water first, the quantity depending on the water content of the fruit (each recipe should tell you). Cooking is done slowly for about an hour until the fruit is very tender. In order to obtain a jelly, the fruit has to be broken down so that the acid and pectin are dissolved in the water.

Use no more than 6 - 8 cups of juice per jelly recipe to ensure proper gel.

30

STRAINING: The easiest method of straining is to use a jelly bag which will drop into a large bowl. They are not very expensive and can be reused. The bag should be scalded before using. If not, use a strainer with three layers of cheesecloth or a clean linen dish cloth and place the strainer over a large bowl. Pour in hot prepared fruit. Tie the cheesecloth or top of bag closed, hang and let drip into bowl (remove colander) until dripping stops. Or tie the cheesecloth or bag to the legs of an upside-down stool with a bowl underneath.

Using Juice Concentrate: Another way is to juice the fruit in a *hot, steam-type juicer.* The advantage to this is that the concentrate can be made, sealed in hot sterilized jars and stored until you have more time to make the jelly. Do not add sugar.

SUGAR: Measure the juice as you transfer it from the bowl to the saucepan. You will need exactly the same number of cups of sugar. The juice is slowly brought to a boil and then the sugar is added. Stir as the sugar dissolves. When it has dissolved, boil as rapidly as possible without stirring.

It should take about 10 minutes to reach setting point, do not boil too long. Jelly is an exact science; cooking too long or not long enough, too little or too much sugar or pectin or acid will cause the jelly to be either too soft or too stiff. Use the same methods as for jam making to determine the set.

Bottling, sealing and storing are all the same as for making jam except that to seal the jelly, it must be either hot or cold, not warm.

'Where Did I Go Wrong?'

JAMS & JELLY FACTS

Bernardin of Canada, Limited

SEAL FAILURE is most often the result of failure to process the finished product for the appropriate time. Other reasons include chipped or cracked jars, failure to follow manufacturer's directions for using specific closures, i.e. boiling lids to soften sealing compound and applying screw bands just until fingertip tight, food particles left on jar rim, and using lids more than once.

MOLD is due to imperfect seals, unsterilized jars and lids, warm and damp storage. Discard jellied products with extensive mold. If a jellied product displays only a very small amount of mold, scrape off the mold plus 1 inch of the product underneath. *If in doubt, throw it out.*

FERMENTATION is due to imperfect seals, inadequate sugar levels, failure to process finished product and improper storage. Microorganisms that cause fermentation survive in the jar, growing over a period of time, fermenting the product. Throw it out.

BUCKLING LIDS are the result of applying screw bands too tightly—no give between lids and screw bands. The build up of pressure inside the jars can be so great that lids buckle or bend out of shape. Overtightening screw bands can also result in jars breaking during processing.

DISCOLORED FRUIT is fruit at the top of a jar that turns brown. Air left in the jar permits oxidation which turns the fruit off-colour. The causes are insufficient syrup covering the fruit or too much head-space left in jar or jars not processed long enough to destroy enzymes.

FLOATING FRUIT is fruit that floats to the top of the jar so that the bottom inch or so shows syrup only. The cause is syrup that is too heavy or packing fruit too loosely in jar or packing raw (unheated) fruit in jar. Hot packing helps to force air out of the tissues of the fruit and will help limit floating, discoloration of fruit and increases the vacuum obtained in a jar and allows you to put more fruit in the jars.

TOO STIFF JAM & JELLY is due to too much pectin in proportion to acid and sugar or cooking no-added-pectin products too long. Nothing can be done for pectin-added preserves. It's not feasible to do them over with more liquid, however, they may still be tastier than store-bought.

TOO SOFT JELLY results from cooling too long, when recipe is doubled and boiling time goes beyond the ideal time limit, cooking too slowly for too long a time, too much sugar; too little sugar or pectin or acid or not cooking long enough. Sometimes you can salvage such jelly by cooking it over.

WEEPING JELLY is partial separation of liquid from other ingredients. This is caused by too much acid, jelling too fast, or storage is too warm. Check seals and make sure there is no mold or fermentation. Move product to a cool, dark, dry place. This should prevent problem from getting worse. Weeping jelly is still usable; just before serving, decant the jelly (pour off liquid).

RUNNY JAM results from undercooking, too little pectin, or improper proportions of fruit and sugar. Jam isn't supposed to be as firm as jelly, so if jam is only a little looser than you'd like it to be, don't bother to remake it. If jam is really thin, try one of the too-soft jelly remedies on one jar. If the test jar doesn't turn out right, make sure all the remaining seals are intact and that storage is in a cool, dark and dry place. Then mark the remaining jars to use as sweet dessert toppings.

REMAKING PROCEDURES

For Jam and Jelly

POWDERED PECTIN: First, using these directions recook a 1 cup trial batch of jelly or jam. If successful, measure total amount of jelly or jam to be recooked. *Do not recook more than 8 cups at one time.*

For each 1 cup jelly or jam, measure 1 tbsp. water and 1½ tsp. powdered pectin. (Stir pectin in package before measuring.) Combine water and pectin in large saucepan, stirring constantly. Bring to a boil. Remove from heat, stir in jelly or jam. For each cup of jelly or jam being recooked, add 2 tbsp. sugar. Bring to a full rolling boil over high heat, stirring constantly. Boil rapidly for 30 seconds. Remove from heat. Skim off foam, and ladle into hot sterilized jars. Seal and process 5 minutes in boiling water.

LIQUID PECTIN: First, using these directions recook a 1 cup trial batch of jelly or jam. If successful, measure total amount of jelly or jam to be recooked. *Do not recook more than 8 cups at one time.*

For each 1 cup jelly or jam, measure and combine 3 tbsp. sugar, 1½ tsp. lemon juice and 1½ tsp. liquid pectin. Place jelly or jam in a saucepan. Bring to a boil, stirring constantly. Add sugar, lemon juice and liquid pectin. Bring to a full rolling boil, stirring constantly. Boil rapidly for 1 minute. Remove from heat. Skim off foam, ladle into hot sterilized jars. Seal and process 5 minutes in boiling water.

LIQUID PECTIN—UNCOOKED (FREEZER): First, using these directions, remake a 1 cup trial batch of jelly or jam. Measure jelly or jam to be remade. *Do not remake more than 8 cups at one time.*

34

For each 1 cup jelly or jam, measure 3 tbsp. sugar and 1½ tsp. lemon juice into a bowl. Add jelly or jam and stir about 3 minutes or until sugar is dissolved. For each cup jelly or jam, add 1½ tsp. liquid pectin; stir about 3 minutes until well blended. Pour into clean containers. Cover with tight lids. Let stand in refrigerator until set. Store in refrigerator or freezer.

POWDERED PECTIN—UNCOOKED (FREEZER): First, using these directions, remake a 1 cup trial batch of jelly or jam. Measure jelly or jam to be remade. Do not remake more than 8 cups at one time.

For each 1 cup jelly or jam, measure 2 tbsp. sugar into a bowl. Add jelly or jam and stir about 3 minutes or until sugar is dissolved. Set aside. In a small saucepan, measure 1 tbsp. water and 1½ tsp. powdered pectin for each 1 cup jelly or jam being remade. Stir and cook over low heat until pectin is dissolved. Add to sugar-jam mixture and stir 2 to 3 minutes or until pectin is thoroughly blended. Pour into clean containers. Cover with tight lids. Let stand in refrigerator until set. Store in refrigerator or freezer.

"Most people are about as happy as they make up their minds to be."

35

PRESERVING BERRIES

There is nothing nicer than fresh berries. When fresh is un-available, there is nothing nicer than preserved berries! Here are some tips on how to preserve that fresh on the farm taste all year 'round by freezing, canning and juicing.

FREEZING BERRIES

All berries can be successfully frozen. There are a few meth-ods for freezing that you can choose from. The most popular and desirable is the IQF method.

INDIVIDUAL QUICK FREEZE (IQF): IQF berries are fast and easy to prepare. The benefits include fresher tasting and firmer berries. IQF allows you to use just one berry or fifteen at a time. Because there is no sugar added, you can use them in any recipe. You can take out a handful to eat or use the entire bag.
Method: Wash and de-stem berries. Sort any undesirable fruit and remove all waste or leaves. Leave fruit whole. Drain well and dry as best as possible. Place fruit *one layer deep* on a cookie sheet. Put in the entire tray in the freezer. Once completely frozen (couple hours), place berries in a heavy duty ziplock bag. Any size will do but remove the air before zipping.

SYRUP: This method requires that you make a syrup with sugar and water according to your taste. *See Syrup Chart.* This method limits the use of your fruit and cannot be used in most recipes or for making jam and jelly later but is nice as a dessert or poured over ice cream.
Method: Wash, stem and dry berries. Prepare syrup. Prepare con-tainers, preferably plastic though glass can be used. Pack fruit with syrup or juice 2 inches from top of jar. Cool. Wipe top of jar clean, place lid on tight. Date and label the container. Freeze.

SUGAR: This method is good if you like your berries sliced or halved. The IQF method is great for those berries that are perfect and therefore good for decorating. However, if the fruit is small, bruised or generally not eye-catching, this method allows you to salvage the most out of the fruit. Thought must be given to the size of container used because you must thaw the entire amount, not just a few berries.

Method: Wash, stem and dry berries. Mix 1 cup sugar with each 4 cups berries in a large bowl until juice is released and sugar is mostly dissolved. Prepare containers. Pack fruit allowing for 2-inch headspace. Seal with lids. Freeze.

UNSWEETENED: Freezing whole or sliced fruit without sugar offers more flexibility than with sugar. This fruit could be used in some recipes if you measure in cups and store in pre-measured containers. An unsweetened solution can be used to enhance the color but this restricts use in recipes. Either way, the fruit will be stuck together upon thawing.

Method: Wash, stem and dry berries. For better colour, prepare a solution of 1 tsp. ascorbic-acid mixed with 4 cups water. Pour over prepared fruit in jars or containers, leaving 2-inch space. Or simply leave out the water solution and pack dry, no sugar. Place lids on and freeze.

TO THAW: Leave fruit in jar or container to thaw in the refrigerator or at room temperature or in a pan of cool water.

To eat raw, do not completely thaw; a few ice crystals improve the texture. *To cook,* thaw until loosened, then cook as a fresh fruit. Some recipes will call for more liquid. Drain the fruit and reserve the fruit juice and use as part of the liquid. It may be necessary to heat the juice first, microwave is easiest.

Refrigerator (fresh): 1 - 2 days
Refrigerator freezer (frozen): 2 - 3 months
Deep Freeze (frozen): 1 - 2 years

CANNING BERRIES

Home canning is one of the easiest and most satisfying ways to ensure having your favorite summertime fruits all year-round. Canning halts spoilage of fresh foods by heating the food in sealed containers. The heat destroys the troublesome organisms and the sealed containers prevent recontamination. Follow the directions carefully and you'll enjoy 'the fruits of your labour' all winter!

AMOUNT TO BUY: Generally buy 1½ to 3 lbs. of berries to can 1 quart.

JARS AND LIDS: Use only standard mason jars (quart jars are more efficient) and the lids designed to fit them (lid and screw band). Discard jars with chips or cracks.

The *flat metal lid* with sealing compound and metal screw band is the most popular type of cap for home canning. This lid is self-sealing, do not tighten the band after processing. Always use a new self-sealing lid. The *rubber ring* type of cap is a three-part process; rubber ring, glass lid and metal screw band. Do not use screw bands that are rusty. They will cause sealing failures.

Wash all jars and glass lids in hot, soapy water and rinse thoroughly or wash in the dishwasher. Boil jars in canner for 15 minutes to sterilize. Boil either the rubber bands or the metal self-sealing lids in water, keep hot until ready to use.

CANNERS: There are two types of canners: *(212°F) boiling water bath* used to sufficiently destroy harmful organisms in high-acid foods such as fruits and a *pressure canner (240°F)* for low-acid foods such as vegetables, meat and soups.

Since you will be using a water bath canner, see *'Making Jam'* chapter under *'Sealing'* section for details.

FILLING THE JARS: Use either raw-pack or hot-pack method:

Raw-Pack: Pack the uncooked fruit into containers. Add boiling syrup, juice or water.

Hot-Pack: Partially cook fruit before packing, then add boiling liquid.

The following steps apply to both raw and hot-pack:

a) Pack prepared fruit into hot jars, leaving specified headspace from top. Shake jars gently.

b) Ladle boiling liquid over the fruit, leaving the specified headspace.

c) Gently work blade of spatula or knife around inside filled jar to eliminate air bubbles. Add more boiling liquid if necessary.

d) Wipe off rim of jar with a damp cloth. Bits of food or syrup on the rim could prevent a perfect seal after processing.

e) Prepare lids. For flat metal lid, place compound side down. Add metal band and screw down until it is firm and tight. For the other lids, place rubber ring on glass lid carefully and place on jar. Add metal screw band until it is firm and tight.

f) Transfer filled jar to canner. Complete filling, covering and placing each jar in canner before filling another.

TESTING THE SEAL: *For metal lids:* press center of lid on cooled jar. If the dip in lid holds (is curved), jar is sealed. *For the glass lids:* tip the cooled jar. If it doesn't leak, jar is sealed. If jar isn't sealed, check for flaws, repack and reprocess with a new lid or rubber ring the full length of time. Or refrigerate and serve within a week or so.

STORING: Remove screw bands if desired. Label and date. Store in cool, dark and dry place.

DETECTING SPOILAGE: Leakage, patches of mold, foamy or murky appearance, bulging lids and off-odor are signs of spoilage. If the canned food doesn't look or smell good, don't use it. See, *'Where Did I Go Wrong?'* section.

JUICING BERRIES

Juicing is the process of extracting juice from whole berries. The extracted juice is called concentrate. No sugar is added. The concentrate can be used fresh or preserved for future use by canning or freezing. Concentrate can be used to make jelly, sauces, or juice.

EQUIPMENT

There are many juicers on the market today. However, when working with berries, the cold juice extractors can be very messy and frustrating to use. The other juicer available on the market is a *hot, steam, aluminum juicer.* The initial investment may be high but overall, making your own juice to use all winter can be very economical and nutritionally better for you. The other method of extracting juice is done with a 'jelly bag' or using a strainer and 3 layers of cheesecloth. See *'Making Jam and Jelly.'*

The following instructions on juicing will pertain to using the *hot, steam, aluminum juicer.*

Use mason pint or quart jars with appropriate lids. See *Canning* section, *'Jars and Lids.'*

FRUIT

Buy fresh, ripe to slightly overripe berries to juice. Quite often berries are on sale if you buy at the source and if they are more than one day old or overripe. Because it takes more berries to make juice than jam, you might want to look for these sales.

The beauty of juicing is that you only need to carefully wash the berries and throw them into the hopper. It is not necessary to stem or destrig the fruit. Once the juice has been extracted, discard the remaining pulp, it is of no use anymore. Repeat procedure if you have more fruit.

Raspberry and Black Currants make excellent juice to drink. Raspberry, gooseberry, blueberry and black and red currants make excellent jelly.

How Much To Buy: (Approximate)
15 lbs. whole berries = 5 quarts concentrate juice
Any amount you buy can be juiced, it does not have to be a specific amount.

BOTTLING: As with making jam and jelly, the concentrate from juice must be preserved in a safe manner. The safest way to ensure a complete seal is to use mason jars, properly boiled lids and follow the directions for *'Canning.'*

PROCEDURE: Using a *hot, steam-type juicer:*
1) Fill bottom pan with water to prescribed water level and place on stove. Bring to boil on large front burner.
2) Prepare fruit. Wash carefully and fill the hopper with the fruit.
3) Place the middle pot that catches the hot juice on top of bottom pan of water. Place the hopper full of washed fruit on top of juice catcher. Place lid on top of all. Make sure the clamp is in place on the rubber hose attached to the bottom pan.
4) While juicer is juicing, prepare jars and lids. Have screw bands ready.
5) Place large catcher bowl on top of stool just underneath the rubber hose where the juice will drain out of when ready.
6) Bottle when there is enough juice.
7) Let jars stand for 12 hours. When cooled, wash outer jars. Check for proper seal.
8) Label, date and store.

41

To Serve Juice:
1 part juice concentrate
4 parts water
Sugar to taste

Mix the juice concentrate in a large pitcher (glass is always nice) with sugar. Stir until sugar is dissolved. *Slowly* pour water in concentrate, stirring constantly. This reduces the amount of foam. Remove any foam. Pour over ice and garnish with a slice of lemon or orange.

Note: Apple juice is nice mixed with raspberry or black currant juice or all three. Raspberry and black currant are a favorite. Mixing any juice concentrate with soda water or ginger ale makes a nice punch. Mixing a juice concentrate with hot water instead of cold water makes a nice alternative to tea or coffee.

"Children have neither past nor future, they enjoy the present; which very few of us do."

APPETIZERS

AMARETTO CREAM

Serve as a fruit dip

1 cup lowfat dairy sour cream   4 tbsp. Amaretto
4 tbsp. powdered sugar

In small bowl, combine cream, sugar and Amaretto. Whisk until well blended. Cover and refrigerate for 1 hour to blend flavors. Use as a dip for berries or serve as a sauce on top of fresh berries in individual glass serving dishes.

Serves 8

ORANGE POPPY SEED DIP

1 cup orange lowfat yogurt   2 tsp. poppy seed
1 tsp. grated orange peel

Combine all ingredients, blend well. Cover and refrigerate until ready to serve. Use as a fruit dip with all your favorite fresh berries and fruit.

Makes 1¾ cups

LORRAINE'S BERRY DIP

Fresh Strawberries   2 cups demerara sugar
2 cups sour cream, room temperature, stirred

Wash and drip dry fresh strawberries leaving the stems on. Put each ingredient in glass bowls. Dip berry in sour cream first, then in the sugar. Pop in mouth...*mm*...*mm*...*good!*

STRAWBERRY CHOCOLATE DIP

4 oz. semi-sweet chocolate
12 whole strawberries, with stems, washed and drained

Chop chocolate squares into 4 pieces each. Partially melt chocolate over hot water until you can still see some small pieces of chocolate in the mixture. (Microwave not recommended) About $^1/_3$ of the chocolate should be unmelted.

Remove from heat and continue stirring until completely smooth. Set the chocolate over a pan of lukewarm water and dip the strawberries. Refrigerate berries on waxed paper to set.

STRAWBERRY (BUTTER) DRESSING

2 cups Strawberries   1 small red onion, sliced

Dressing:

$^1/_3$ cup vegetable oil   ½ tsp. salt
3 tbsp. cider vinegar   ½ tsp. paprika
2 tbsp. water   ¼ tsp. pepper
1½ tbsp. liquid honey   1 tbsp. poppy seeds

Whisk together all dressing ingredients.

Suggested Salad: Best served on individual plates of butter lettuce topped with red onion slices and fresh strawberries, halved.

Serves 4

"When you are a mother, you are never really alone in your thoughts...A mother always has to think twice, once for herself and once for her child."

STRAWBERRY (SPINACH) DRESSING

Dorothy Miller, Langley

2 bunches spinach leaves, washed and torn
2 cups Strawberries, halved
Dressing:

½ cup sugar
2 tbsp. sesame seeds
1 tbsp. poppy seeds
½ minced red onion

¼ cup Worcestershire Sauce
¼ tsp. paprika
½ cup vegetable oil
¼ cup cider vinegar

Toss spinach and strawberries together. Prepare dressing and mix well. Pour over salad, toss and serve immediately.

Serves 10 - 12.

STRAWBERRY DRESSING

½ cup fresh strawberries
2 tbsp. vegetable oil

3 tbsp. lemon juice
1 tsp. coarse-grain mustard

Press strawberries through nylon sieve into a bowl. Stir in the remaining ingredients. Drizzle over salad (red lettuce, chicory).

FRESH RASPBERRY VINAIGRETTE

¼ cup raspberry vinegar
1 cup light olive oil

1 tbsp. lemon juice
¼ puréed red raspberries

Whisk the vinegar into the oil, with the lemon juice and salt and pepper. Stir in raspberry purée.

Enough for 6 salads.

POPPY SEED DRESSING

for Fresh Fruit Salad

½ cup honey   1 tsp. grated orange peel
4 tbsp. frozen orange conc.   ½ tsp. poppy seed
4 tbsp. oil   ¼ tsp. dry mustard

In small jar with tight-fitting lid, combine all ingredients. Shake well. Refrigerate 2 hours to blend flavors. Arrange fruit that is cleaned, peeled and sliced on individual plates. Drizzle dressing on top. Serve with garnish of fresh picked flowers.

Serves 8

STRAWBERRY & AVOCADO SALAD

2 cups fresh sliced or halved   2 small ripe avocados
strawberries: washed, drained   Fresh mint or strawberry
and hulled   leaves for garnish

Honey Lemon Dressing:

2 tbsp. sunflower oil   ¼ tsp. paprika
2 tsp. honey   Salt and pepper to taste
2 tbsp. lemon juice

Mix together all the ingredients for the honey lemon dressing. Just before serving, pit, peel and dice avocado. Arrange decoratively the diced avocado on top of the strawberries on 4 plates. Drizzle with dressing, garnish with leaves.

"The persons hardest to convince they're at the retirement age are children at bedtime!"

AVOCADO RING WITH STRAWBERRIES

2 boxes (3 oz. each) lime or
    lemon flavored gelatin
1½ cups cold water
2 very ripe avocados, peeled,
    pitted and mashed
Salad greens

½ tsp. salt
2 cups hot water
2 tbsp. fresh lemon juice
⅓ cup mayonnaise
3 cups Strawberries, sliced
    (reserve ½ cup for dressing)

Dissolve gelatin and salt in hot water. Add cold water and chill until slightly thickened. Pour lemon juice over avocados. Stir avocados and mayonnaise into gelatin, blending well. Pour into 5-cup ring mold and chill until firm. Unmold ring on greens and fill with 2½ cups of the berries. Serve with Honey Cream Dressing.

Honey Cream Dressing: Mix ½ cup *each* of dairy sour cream and mayonnaise, 1 tbsp. honey, and ½ cup sliced berries.

Serves 8.

FRESH BLUEBERRY AND LIME MOLD

An alternative to everyday salads!

2 pkg. unflavored gelatin
1 cup cold water
1½ cups hot water
½ cup sugar
¼ tsp. salt

½ cup fresh lime juice
3 cups fresh blueberries
Salad Greens
Mayonnaise

Soften gelatin in cold water. Let stand 5 minutes. Add hot water, sugar and salt. Mix well to dissolve gelatin. Blend in lime juice. Chill until mixture is about as thick as fresh egg whites. Fold in blueberries. Turn into a 5-cup mold and chill until firm. Turn out on top of salad greens on serving plate. Serve with mayonnaise. Makes 6 - 8 servings.

STRAWBERRY DREAM SALAD

1 tbsp. or ¼ oz. envelope
  unflavored gelatin
2 cups strawberry-flavored
  yogurt

2 cups fresh strawberries,
  halved or sliced if large
3 tbsp. water

In a small bowl, sprinkle gelatin over water. Let soak until soft-ened. Place the bowl over a small saucepan of hot water, stirring until dissolved. Remove from heat and cool slightly.

Stir in a little yogurt, then gradually whisk in remaining yogurt. Pour into a wetted mold or 4 wetted individual molds. Refrigerate to set. To serve, turn out mold, or molds onto serving plate or plates. Surround with sliced strawberries.

STRAWBERRY SALAD SUPREME

2 envelopes unflavored gelatin
1 cup reconstituted frozen
  lemonade
2 cups ginger ale
¾ cup mayonnaise
8 maraschino cherries, halved

¼ cup maraschino cherry
  liquid
¼ tsp. salt
1 cup strawberries, sliced
¼ cup slivered almonds

Sprinkle gelatin over lemonade in small saucepan to soften. Heat over low heat, stirring constantly until gelatin is dissolved. Pour into large bowl; add ginger ale, mayonnaise, cherry liquid and salt. Beat with wire whisk or beater until smooth. Chill until mixture is consistency of unbeaten egg white. Fold in strawberries, almonds and cherries. Pour into 5-cup mold and refrigerate until firm. Unmold on serving plate; garnish as desired.

Makes 6 - 8 servings.

BERRY MEDLEY SOUP

1½ lbs. fresh or frozen mixed berries (raspberries, red currants, black currants)
¼ cup sugar
1 cup sweet white wine
1 - 4 inch cinnamon stick
2 cups water
½ cup whipping cream, lightly whipped to serve

Reserve a few raspberries for garnish. In a large saucepan, mix remaining berries, sugar, wine, cinnamon and water. Place over low heat, stir with wooden spoon until sugar dissolves. Simmer until berries are soft, about 5 to 10 minutes.

Discard cinnamon stick. Press mixture through a nylon sieve. Cover and refrigerate at least 1 hour. To serve, swirl cream into soup and garnish with reserved raspberries.

ICED STRAWBERRY SOUP

2 cups fresh strawberries, washed, hulled and dried
1 medium English cucumber, peeled and cut in chunks
2 cups lemon yogurt
Thin lemon slices
½ cup light rum
3 tbsp. chopped fresh mint leaves
2 tbsp. sugar
Mint sprigs

Purée strawberries in a food processor. Pour into a large bowl and set aside. Place cucumber chunks, yogurt, rum, chopped mint leaves and sugar in food processor. Process until cucumber is fully puréed. Pour into bowl of puréed strawberries and stir well. Chill in refrigerator. Serve ice-cold, garnished with sprigs of mint and lemon slices.

Makes 4 cups.

"If we can't keep up with our children, we can at least get behind them."

49

NORWEGIAN BLUEBERRY SOUP

1 envelope unflavored gelatin
2 cups fresh blueberries,
  washed
4 cups fresh orange juice

¼ cup sugar
¼ cup cold water
Fresh mint
3 tbsp. fresh lemon juice

Soften gelatin in cold water in a custard cup. Place in a pan of hot, not boiling, water until melted and ready to use. Combine orange and lemon juices, sugar, and melted gelatin. Stir until sugar and gelatin are dissolved. Chill until mixture begins to thicken. Fold blueberries into mixture. Chill until ready to serve. Spoon into chilled bouillon cups. Garnish with fresh mint.

Serves 6 - 8.

STRAWBERRY BUTTERMILK SOUP

5 cups strawberries, hulled
2 cups water
½ cup buttermilk
½ cup freshly squeezed
  orange juice
Juice of 2 lemons

1 tbsp. chopped
  fresh mint
¼ cup brown sugar
4 - 6 tbsp. plain yogurt
Mint sprigs for garnish

In a stainless steel or enameled saucepan, combine strawberries, water, buttermilk, orange juice, lemon juice, chopped mint and sugar. Bring to a boil, reduce heat and simmer gently for 10 minutes. Cool slightly, purée in blender or food processor and chill thoroughly. Before serving, swirl one tablespoon yogurt through each portion and garnish with mint sprigs.

Serves 4 - 6.

"Our children are likely to live up to what we believe in them."

CHILLED CURRANT SOUP

¾ cup fresh red currants
¾ cup fresh black currants
¾ cup fresh/frozen cranberries
1½ cups medium-dry white
    wine
½ cup sugar
1 2-inch cinnamon stick

1 orange; finely grated peel
    and juice
1¼ cups water
1 tbsp. creme de cassis liqueur
²/₃ cup dairy sour cream
Black currant leaves to
    decorate, if desired

In a large saucepan, bring currants, cranberries, wine, sugar, cinnamon, orange peel, juice and water to a boil. Lower heat and cook gently until fruit is tender, about 15 minutes. Discard cinnamon stick. In a blender or food processor, purée fruit and liquid. Press through a nylon sieve to remove seeds. Cool, then refrigerate 1½ hours.

Stir creme de cassis into soup, then pour into chilled soup bowls. Carefully place a dollop of sour cream on each serving. Using a skewer, feather the sour cream attractively. Decorate with black currant leaves, if desired. Serve immediately.

Serves 4 - 6.

"Words have incredible power to build us emotionally. Many of us can clearly remember words of praise our parents spoke years ago."

MAIN COURSES

RASPBERRY CHICKEN

Kathy Stockner, Hazelton, B.C.

4 chicken breasts	1 clove garlic, crushed
¼ tsp. salt	1 tbsp. minced parsley
⅛ tsp. pepper	¼ cup chicken stock
2 tbsp. butter	1 tsp. crushed green
2 tbsp. oil	peppercorns
1 pint fresh raspberries	2 tbsp. butter, cut into pieces
3 tbsp. white or rosé wine	sliced, sautéed mushrooms
	(optional)

Sprinkle chicken with salt and pepper. Heat butter and oil in large frying pan. Brown chicken on all sides. Mash half the berries through a sieve to remove seeds. Refrigerate the remainder.

In a small bowl, combine mashed berries, wine, garlic, parsley and stock. Pour over chicken, cover and cook for 15 minutes or until chicken is done. Remove chicken to serving platter and keep warm.

Stir crushed peppercorns into sauce. Remove pan from heat and stir in butter pieces until they melt and sauce is well blended. Pour sauce over hot chicken and garnish with reserved raspberries and mushrooms. Serve immediately.

Serves 4.

"A father is a fellow who has replaced the currency in his wallet with snapshots of his kids."

BLUEBERRY SALSA

(Great for BBQ Salmon)

½ large pink grapefruit
1 jalapeño pepper, chopped
1 tsp. honey
1 cup blueberries, fresh or
   thawed

2 tbsp. red onion, finely
   chopped
1 tbsp. lime juice
2 tbsp. chopped fresh cilantro

Section pink grapefruit and discard membrane. Dice grapefruit and mix with other salsa ingredients.

To Serve: Prepare the barbecue and oil the grill well. Cut the salmon into 4 filets (1½ lb. each). Barbecue skin-side down over medium high heat for about 10 minutes per inch of thickness. Spoon salsa over the salmon or serve on the side.

STRAWBERRY MINT SALSA

Great served over halibut steaks

2 cups strawberries diced
¼ cup red onion, chopped
2 tbsp. fresh lime juice

2 tbsp. orange juice
2 tbsp. fresh mint, chop.
Fresh mint sprigs, garnish

Combine all ingredients in a glass bowl and marinate in the refrigerator for 1 to 2 hours.

Suggested Serving: Arrange 4 (6-7 oz) halibut steaks brushed with olive oil and lightly seasoned with salt and fresh pepper (baked at 400°F for 15-20 min.) on plates. Spoon salsa over fish and garnish with mint sprigs.

"We must do the best we can with the best of what we have"

FRESH RASPBERRY SALSA

Serve with grilled or broiled halibut.

3 cups fresh raspberries,
    divided
½ tomato
1½ tbsp. cilantro
½ med. red onion
Ground cumin powder

3 green onion, sliced
¼ jalapeño pepper
1 tbsp. olive oil
1 tsp. sugar
Salt and fresh pepper

Finely chop the tomato, cilantro, red onion and jalapeño pepper. Purée two cups raspberries in food processor. Strain to remove seeds. Place purée in bowl with remaining ingredients except whole raspberries. Mix thoroughly. Gently fold in raspberries. Chill.

Makes 2 cups.

GOOSEBERRY SAUCE

Good side dish for fish, goose, duck and roast pork.

2 cups gooseberries
Handful of spinach or sorrel
    leaves, chopped (optional)
Sugar to taste

1 cup water
1 tbsp. butter
⅛ tsp. grated nutmeg

Cook gooseberries and water together until berries are soft. Add spinach leaves; cook 3 minutes longer. Drain and reserve juice. Rub berries and spinach through a food mill or a fine sieve. Return to saucepan and combine with juice. The purée should be of the consistency of a thin batter. If too liquid, boil over high heat until right consistency is achieved. Add butter, nutmeg, and sugar to taste. Simmer for 2 to 3 minutes, stirring constantly. (If no spinach is used, add a few drops of green food coloring for right color.)

Makes 1½ cups.

RED CURRANT GLAZE

Use this glaze for basting on pork, in the oven or on the BBQ
—delicious!

½ cup red currant jelly   2 tbsp. honey
2 tbsp. lemon juice

Combine all ingredients in a small saucepan. Stir over low heat until jelly melts. Baste on pork during last 30 minutes of cooking.

Pork Loin: Rub 4 lb. of pork with a mixture of 2 tbsp. lemon juice, 3 tbsp. olive oil, 1 tsp. rosemary, 1 crushed garlic clove and salt. Marinate one hour. Roast at 350°F about 35 min./lb. or until meat is tender and juices run clear. During last 30 minutes, baste with glaze.

CASSIS SAUCE

Served over Barbecued Duck Breast

2 cups fresh black currants   1 tbsp. red currant jelly
¼ cup water   1 cup duck stock
2 tbsp. sugar   2 tbsp. creme de cassis
¼ cup red wine vinegar   ¹/₃ cup butter, chilled

In a small sauce pan, simmer black currants with water and sugar until tender. Combine vinegar and jelly in another pan, reducing over medium heat to a syrup. Add cassis and scrape up contents. Stir in stock and heat through. Pour onto black currants and simmer 20 minutes. Strain into a clean saucepan and taste for flavor. Reheat just before serving, swirling in butter piece by piece.

"We need time alone to discover who we really are."

RED RASPBERRY VINAIGRETTE

(Best served over veal)

½ cup olive oil   2 tbsp. raspberry concentrate
6 tbsp. raspberry vinegar

Whisk together olive oil, raspberry vinegar and concentrate. Drizzle over thinly sliced veal, warm or chilled on individual serving plates. Add fresh mint leaves and fresh raspberries to decorate each plate.

STRAWBERRY BARBECUE SAUCE

Pamela Allardice

1¼ cups strawberry jam   Freshly ground black pepper,
¾ cup tomato sauce   to taste
1 tbsp. Worcestershire sauce   1 tbsp. honey
1 onion, chopped fine   1 tsp. finely chopped fresh
     ginger

Combine all ingredients in stainless steel or enameled saucepan; bring to a boil, stirring constantly. Reduce heat and simmer for five minutes, adding a little water if sauce becomes too thick. Brush on chicken or meat.

Makes 2 cups.

GOOSEBERRY CHUTNEY

1 cup chopped onions   4 lbs. gooseberries, trimmed
1½ cups raisins   and washed
1½ lbs. sugar   1 tbsp. salt
1 tsp. cayenne pepper   2¼ cups vinegar
2 tsp. ginger

continued next page

(Goodeberry Chutney) ...Boil the onions in a little water until they are soft, then drain off the water. Place the onions with the other ingredients into a preserving pan. Bring to a boil then reduce the heat and simmer, stirring frequently, until the chutney is thick. Pour into hot sterilized jars. Seal, process and store.

GOOSEBERRY CURD

2 lbs. gooseberries, trimmed and washed
3 eggs slightly beaten

2 oz. water
$^1/_8$ lb. butter
1 lb. sugar

In a large saucepan, bring the gooseberries and the water to a boil over high heat. Reduce the heat to low, cover the pan and simmer for 20 to 25 minutes, or until the gooseberries are soft and mushy.

Remove the pan from the heat. Push the fruit through a strainer into a medium-sized mixing bowl, pressing down on the fruit with a wooden spoon. Discard the skins left in the strainer. In another bowl set over a pan of simmering water, melt the butter. Stir in the eggs, sugar and the gooseberry purée. Cook the mixture, stirring frequently, for 25 to 30 minutes, or until it thickens. Remove the bowl from the heat and pour curd into sterilized jars. Seal, process and store.

CANADIAN SPICED GOOSEBERRIES

This makes a fine relish for meat or fowl.

4 lbs. gooseberries
2 cups firmly packed light brown sugar
½ cup cider vinegar

¼ tsp. salt
¼ tsp. *each* of ground allspice, cinnamon and cloves

continued next page

57

(Canadian Spiced Gooseberries) …Wash berries, and remove blossoms and stems. Discard any soft berries. Combine first 4 ingredients in a large saucepan. Mix well. Cook over medium heat for 30 to 40 minutes, or until thickened. Stir frequently. Add spices about 10 minutes before cooking time is up. Pack in hot sterilized jars. Seal, process and store.

Makes 4 - ½ pint jars.

BLUEBERRY RELISH & SYRUP

3 ½ cups frozen blueberries
1 apple, peeled and diced
½ cup sugar
1 cinnamon stick

1 tsp. whole cloves
2 tsp. cider vinegar
4 tsp. fresh lemon juice
½ tsp. aromatic bitters

In a large saucepan, combine blueberries, apples, sugar, cinnamon and cloves. Bring to a boil, reduce heat and simmer for 3 minutes. Add vinegar, lemon juice and bitters. Stir carefully just to mix; remove from heat and let cool.

Pour through large strainer to drain off juice. Reserve syrup; use to baste and glaze a small ham during baking. Discard cinnamon and cloves. Chill remaining mixture for relish.

"Children aren't ever sure of what they want, but that's not the point; it's that they want it now!"

DESSERTS

Cakes, Cheesecakes, Shortcakes, Muffins

ANNA'S FESTIVE SUMMER FRESH CAKE

Anne-Marie Driediger

"I like to arrange the fruit in new designs—its fun! This cake is fast, delicious and very colorful, a hit for any occasion."

2 cups all-purpose flour	1½ cups sugar
3 tsp. baking powder	1 tsp. salt
1 cup milk, room temperature	½ cup soft shortening
1 tsp. vanilla or almond extract	2 eggs, room temperature
Fresh fruit, washed and dried	Cream cheese frosting

Preheat oven to 350°F. Grease and flour one 13x9-inch glass or two 8-inch round cake pans. In a large bowl, mix flour with sugar, baking powder and salt. Add shortening, mix well. Add milk and vanilla, mix well. When smooth and creamy, add eggs, beat 2 minutes longer. Pour batter into prepared pans. Bake round pans for 30 to 35 minutes, oblong pan for 35 to 40 minutes. Cool in pans for 10 minutes. Remove from pans. Cool thoroughly on wire racks. Frost with Cream Cheese Icing (see below). Wash and thoroughly dry fresh fruit of your choice. Arrange the fruit. Strawberries, raspberries and blueberries make a wonderful contrast together.

Cream Cheese Frosting:

1 - 8 oz. pkg. cream cheese,	4 tbsp. soft butter
softened at room temperature	3 cups icing sugar
3 tsp. vanilla or almond extract	

continued next page

59

(Anna's Festive Summer Fresh Cake) Blend cream cheese and butter until smooth. Add extract and icing sugar gradually. Blend until smooth and creamy. Makes enough for two 13x9-inch pans spread thin, or one, if spread on thick. If topping with fresh fruit (whole, so fruit doesn't run), use all the frosting and spread thick so fruit can settle in firmly and not slide off cake.

KENTUCKY JAM CAKE

1 cup butter, softened	½ tsp. salt
2 cups sugar	½ tsp. cloves
5 eggs	½ tsp. allspice
1 cup blackberry jam, seedless	1 cup buttermilk
3 cups all-purpose flour	1 cup chopped nuts
1 tsp. baking soda	1 cup chopped raisins
Caramel Frosting	Fresh fruit for garnish

Cream butter and sugar until light. Add eggs, one at a time, beating well after each addition. Add jam, beat well. Add sifted flour, soda, salt, cloves and allspice alternately with buttermilk, beating until smooth, beginning and ending with dry ingredients. Stir in nuts and raisins. Pour into 4 - 9-inch layer greased pans. Bake at 325°F for 30 - 35 minutes. Cool. Spread Caramel frosting between layers and on top of cake. Garnish with fresh fruit.

Caramel Frosting:

2 cups brown sugar, packed	Dash of salt
1 cup sugar	²/₃ cup cream
2 tbsp. light corn syrup	1 tsp. vanilla
3 tbsp. butter	

Place all ingredients in a saucepan. Bring to a boil, cover and cook for 3 minutes. Uncover and cook to 236°F on a candy thermometer or until a small amount of mixture forms a soft ball when dropped into cold water. Cool for 5 minutes, then beat until thick. If too stiff, add a little hot water.

"The best thing parents can spend on their children is time, not money."

STRAWBERRIES-AND-CREAM CAKE

Lynda Brown, Langley

Cake:

6 eggs	1 cup granulated sugar
1 tsp. vanilla	1 cup all-purpose flour
½ tsp. baking powder	Pinch of salt
⅓ cup butter, melted	

Filling:

2 cups whipping cream	2 tbsp. granulated sugar
½ tsp. vanilla	2 cups strawberries, sliced

Cake:

Grease and flour two 9-inch round cake pans. In warmed mixing bowl, beat eggs at high speed until foamy. Gradually beat in sugar, then continue to beat at high speed for 8 to 10 minutes or until batter is thick and pale yellow and falls in ribbons when beaters are lifted. Beat in vanilla.

Sift together flour, baking powder and salt. Sift ⅓ of flour mixture over egg mixture and fold in; repeat twice. Fold in melted butter. Pour into prepared pans. Bake in 325°F oven for 25 to 30 minutes or until cake springs back when lightly touched in center. Let cool for 5 minutes. Turn out onto wire rack and let cool completely.

Filling:

Beat cream until soft peaks form. Beat in sugar and vanilla until stiff peaks form. Split each cake layer in half horizontally. Spread ¼ of filling over cake and sprinkle with ¼ of strawberries. Repeat with remaining layers. Serve immediately or refrigerate for up to 2 hours.

"Treat your children like you would your best friend."

61

BLUEBERRY COFFEE CAKE

2 cups flour
1 tbsp. baking powder
½ tsp. salt
1 cup milk
2 cups fresh blueberries

½ cup sugar
1 tsp. ground cinnamon
½ cup butter, softened
1 egg
Nut Topping

Combine all ingredients except blueberries and nut topping. Using electric mixer, beat 30 seconds on low speed. Beat 2 minutes on medium speed; scrape bowl frequently. Spread half of batter in greased 2-quart glass baking dish. Spread half of blueberries on batter; sprinkle with ½ of nut topping. Repeat layers with remaining, batter, blueberries and nut topping. Bake at 375°F for 45 to 50 minutes or until wooden pick inserted near center comes out clean.

Nut topping: Combine 3 tbsp. butter or margarine, ⅓ cup each packed brown sugar and flour and 1 tsp. ground cinnamon. Add ⅓ cup finely chopped pecans; mix well.

BLUE MONDAY CAKE

Cake:
½ cup sugar
1½ cups flour
2½ tsp. baking powder
½ tsp. salt
1 egg
½ cup milk
¼ cup melted shortening

Crumb Mixture:
2½ cups blueberries
½ cup sugar
⅓ cup flour
¼ cup butter
¾ tsp. cinnamon

Sift together dry cake ingredients. Make hollow and add egg beaten with milk and shortening. Stir just until moistened. Spread in 9x13-inch pan. Cover with blueberries. Make crumb mixture with other ingredients. Sprinkle over berries. Bake at 350°F for 45 minutes.

FRESH BLUEBERRY CAKE

1½ cups sifted
  all-purpose flour
½ tsp. salt
½ cup shortening
²/₃ cup milk
2 tbsp. all-purpose flour
1 tbsp. sugar

2 tsp. baking powder
1 cup sugar
1 tsp. vanilla extract
1 egg
1½ cups fresh blueberries,
  washed
Lemon sauce

Sift flour with baking powder and salt. Gradually blend sugar with shortening and vanilla. Beat in egg. Add flour mixture alternately with milk. Combine blueberries, 2 tbsp. flour, and 1 tbsp. sugar and fold into the batter. Turn into a well-greased lightly floured 9-inch cake pan.

Bake in preheated 350°F oven for 1 hour, or until cake tester or toothpick inserted in center comes out clean. Turn out on wire rack. Serve warm or cold, cut into squares, with lemon sauce. See the section on sauces for Lemon Sauce.

BLUEBERRY COTTAGE-CHEESE CAKE

1 tbsp. soft butter
2 envelopes unflavored gelatin
¼ tsp. salt
¾ cup milk
3 cups cottage cheese, sieved
1 tsp. vanilla extract

4 cups blueberries, washed
¾ cup sugar
1 egg, separated
1 tsp. grated lemon rind
2 tbsp. fresh lemon juice
¾ cup heavy cream,
  whipped

Spread butter on bottom and sides of shallow 1½-quart dish. Arrange 3½ cups berries in dish to form a shell. In top part of double boiler, mix gelatin, sugar and salt. Beat egg yolk and milk; add to gelatin mixture.

continued next page

(Blueberry Cottage-Cheese Cake)…Cook over boiling water, stirring, until gelatin is dissolved, about 6 minutes. Add lemon rind, and cool. Stir in cottage cheese, lemon juice and vanilla extract. Chill until slightly thickened to the consistency of unbeaten egg whites. Fold in stiffly beaten egg white and cream. Pour into shell; put ½ cup berries in center; chill until firm.

Serves 8 - 10.

STRAWBERRY CHEESECAKE

3 pkg. (8 oz. each) cream
   cheese, softened
1 cup crushed strawberries
3 tsp. vanilla, divided
Fresh strawberry halves
   or slices for garnish

¼ tsp. salt
4 eggs
1 cup sour cream* or
   crème fraîche
* Do not use calorie-reduced
   sour cream.

Preheat oven to 325°F. Beat cream cheese in large bowl until creamy. Blend in crushed fruit, 1 tsp. vanilla and salt. Add eggs, one at a time, beating well after each addition. Pour into greased 9-inch springform pan. Bake 50 minutes.

Combine sour cream and remaining 2 tsp. of vanilla; mix well. Carefully spoon over warm cheesecake. Return to oven; continue baking 10 minutes or just until set. Turn oven off; leave cheesecake in oven for 30 minutes with door closed. Transfer to wire rack; loosen cheesecake from rim of pan. Cool completely before removing rim. Cover and chill at least 6 hours or overnight. Just before serving, garnish cheesecake with strawberries.

Makes 10 servings.

"A mother's love perceives no impossibilities."

BERRY CHEESECAKE

Anne-Marie Driediger

Crust:
1½ cups graham crumbs
½ tsp. cinnamon(optional)
½ cup butter, melted

Filling:
2 *each* 250g cream cheese
1 cup sugar
3 eggs, beaten
3 cups sour cream

Glaze:
2 tbsp. lemon juice
3 cups fresh berries, divided
½ cup water
⅓ cup sugar
2 - 3 tbsp. cornstarch

Crust: Combine all ingredients and spread in bottom of 9-inch springform pan. Bake at 350°F for 10 minutes. Cool.

Filling: Cream cheese until soft. Add sugar, eggs. Mix well. Stir in sour cream and lemon juice, blend well. Pour onto crust and bake at 350°F for 40 minutes. Cool and refrigerate until ready to use. Just before serving, remove ring with sharp knife scraping the sides.

Glaze: Use 2 cups of either strawberries, raspberries, blueberries, blackberries or a combination. Combine with water in saucepan and bring to a boil. Simmer gently for 1 minute. Mix sugar and cornstarch together and stir into simmering berries. Cook, stirring constantly until mixture thickens and becomes clear. Stir in remaining berries and cool.

To Serve: Spread enough glaze to cover top of cheesecake. Remaining glaze can be served as extra in a glass bowl. Serves 12.

"To keep children on your team, you must assure them from the beginning that you are on theirs."

OLD-FASHIONED CHEESECAKE

Use any fresh berries, especially currants.

½ cup + 1 tsp. butter, melted
1 cup digestive biscuits, crushed
1 tsp. cinnamon
2 - 250g cream cheese

¼ cup castor sugar
2 ⅓ cup double cream
1¼ lb. fresh berries
½ oz. gelatin, dissolved
1 egg white, stiffly beaten in 2 tbsp. boiling water

Lightly grease a 9-inch springform pan with butter. Combine the biscuits, ½ cup melted butter and cinnamon. Mix well and line the base of the pan, pressing firmly.

Beat cream cheese and sugar until smooth and creamy. Stir in ½ cup cream and 1 lb. of washed and rinsed berries. Beat in the dissolved gelatin and spoon it on the biscuit base. Chill for 30 minutes or until set.

Beat remaining double cream until it forms stiff peaks. Fold beaten egg white into cream. Spoon the cream mixture onto chilled cheesecake, making swirling patterns with the back of a metal spoon. Sprinkle remaining berries over the cream. (Make sure berries are hulled and patted dry.)

Note: This cheesecake freezes well up to 3 months if left in pan and wrapped securely. Seal and label. Thaw overnight in refrigerator. Serve with extra whipped cream!

Makes 6 - 8 servings.

"When a child is allowed to do absolutely as he pleases, it will not be long until nothing pleases him."

BERRY SHORTCAKES

1¾ cups all-purpose flour
1 tbsp. baking powder
⅛ tsp. salt
½ cup cold butter
½ cup milk
1 tsp. vanilla
1 egg
1 tsp. water

1 cup sliced strawberries
1 cup raspberries
1 cup blueberries
3 tbsp. no-sugar-added
 strawberry spreadable fruit
4 tbsp. almond-flavored
 liqueur, divided
1 cup heavy cream

Preheat oven to 425°F. Combine flour, baking powder and salt in medium bowl. Cut in butter with pastry blender or two knives until mixture resembles coarse crumbs. Add milk and vanilla; mix just until dry ingredients are moistened. Knead dough gently on lightly floured surface ten times. Roll or pat out to ½ inch thickness. Cut with 3" heart or round-shaped biscuit cutter; place on ungreased cookie sheet. If necessary, reroll scraps of dough in order to make six shortcakes. Beat together egg and water; brush lightly over dough. Bake 12 to 14 minutes or until golden brown. Cool slightly on wire rack.

While shortcakes are baking, combine berries, spreadable fruit and 3 tbsp. liqueur; let stand at room temperature for 15 minutes. Beat cream with remaining 1 tbsp. liqueur until soft peaks form. Split warm shortcakes, fill with about ⅔ of berry and whipped cream mixtures. Replace tops of shortcakes; top with remaining berry and whipped cream mixtures.

Makes 6 servings.

"If only we could get a little older a little later, and a little wiser a little younger."

OUR FAMILY'S FAVORITE STRAWBERRY SHORTCAKE

June Driediger

White Cake Base:

1 cup sugar	½ cup butter
2 eggs, well beaten	2 tsp. baking powder
Pinch of salt	1½ cups cake flour
¾ cup milk	

Topping:

2 cups sliced strawberries	Sugar to taste
Whipped cream	Vanilla ice cream

Cream butter and sugar. Add beaten eggs. Sift together flour, baking powder and salt. Add milk and flour mixture alternately to cream mixture. Bake in a greased and floured pan at 350°F for 20 minutes or until toothpick comes out clean. Remove from oven and cool on rack. Cut into squares. Slice squares horizontally and fill with ice cream. Sprinkle sugar over strawberries in bowl and let stand 5 minutes. Top squares with berries and ample whipped cream.

"It is important to give our children two things—the first is roots; the other is wings."

68

GRANDMA'S SCONES

Sandra (Evans) Koven, N. Vancouver
"This recipe was handed down to me by my Grandmother,
Grace (Miller) Evans."

3 cups flour	1 tsp. salt
¾ cups sugar	1 egg, slightly beaten
¾ cups butter, softened	Milk
3 tsp. baking powder	1 cup blueberries

Sift together dry ingredients. Do not overmix, hand mixing may give a fluffier texture. Beat the egg in a measuring cup. Add enough milk to make 1 cup. Stir this into batter. Fold in fresh or frozen blueberries, thoroughly drained. Form batter into desired shape and thickness. Bake at 425°F for 10 to 12 minutes.

BLUEBERRY MUFFINS

Joan Gardner, Aldergrove

½ cup butter	2 tsp. baking powder
1 cup sugar	½ tsp. salt
2 large eggs	½ cup milk
1 tsp. vanilla	2 cups fresh/frozen
2 cups flour	blueberries (thawed and
	drained)

Beat butter until fluffy. Add sugar and mix well. Stir in eggs and vanilla. Mix (in another bowl) flour, baking powder and salt. Add half the flour mixture to butter mixture and mix well. Stir in half of the milk, balance of flour and remaining milk. Add the blueberries. Fill each greased muffin cup to the top with mixture. Bake at 375°F for 25 - 30 minutes until muffins are golden. Cool in tins.

Makes 12 muffins.

OLD FASHIONED BLUEBERRY MUFFINS

Blueberry Growers Association

2 cups flour
2 tsp. baking powder
1 tsp. ground cinnamon
¼ tsp. salt
2 eggs

1 cup milk
¾ cup sugar
½ cup vegetable oil
1 cup fresh or frozen
    blueberries, thawed if needed

Combine flour, baking powder, cinnamon and salt. Mix well. Beat in separate bowl, eggs lightly. Stir in milk, sugar and oil. Quickly combine egg mixture into dry mixture. Carefully stir in blueberries. Spoon into greased muffin cups. Bake at 400°F for 15 to 17 minutes.

Makes 18 muffins.

YOGURT BLUEBERRY MUFFINS

Eileen Ferraro, Abbotsford

2 cups yogurt
2 tsp. baking soda
1½ cups brown sugar
2 eggs
1 cup oil

2 cups natural bran
2 tsp. vanilla
2 cups flour
4 tsp. salt
1 cup frozen blueberries

Mix yogurt and 1 tsp. baking soda in separate bowl. Set aside. Beat sugar, eggs and oil together. Add bran and vanilla. Mix flour, 1 tsp. baking soda and salt. Add to sugar mixture alternately with yogurt. Fold in blueberries. Pour into greased large size muffin tins. Bake at 350°F for 30 minutes.

Makes 24 muffins.

"The only thing children wear out faster than shoes are parents and teachers."

Crisps, Crumbles, Cobblers, Buckles

RASPBERRY CRISP IN MICROWAVE

B.C. Raspberry Growers Association

1 - 16 oz. package of frozen
  raspberries
or 3 cups fresh raspberries
  ½ cup flour
²/₃ cup quick cooking rolled oats

2 tbsp. lemon juice
¾ cup brown sugar
¹/₃ cup butter, softened
1 tsp. cinnamon

Spread raspberries in 8-inch microwave-safe baking dish. Sprinkle with lemon juice. Mix sugar, flour, oats, butter and cinnamon. Sprinkle on top of berries. Microwave uncovered on high until raspberries are hot and bubbly, 7 to 10 minutes. Serve warm or cold with whipped cream or ice cream.

LIGHT PEACH AND BERRY CRISP

Nancy Birchall, Vancouver.

3 cups sliced peaches
2 tbsp. flour
1 tbsp. granulated sugar
¼ cup packed brown sugar
1 tsp. cinnamon

2 cups raspberries or
  blueberries
¹/₃ cup rolled oats
2 tbsp. oat bran
¼ cup butter

In 8-inch square baking dish, combine peaches and berries. Mix together sugar and flour; stir into fruit. Combine rolled oats, brown sugar, oat bran, remaining flour and cinnamon. Cut in butter until crumbly. Sprinkle over fruit. Bake in 375°F oven for 30 to 35 minutes or until peaches are tender and topping is browned and crisp.

"Learning to love yourself opens the way for others to do the same."

71

DOUBLE-CRUNCH BUMBLEBERRY CRISP

Crust:

1 cup all-purpose flour	1 cup rolled oats
¾ cup packed brown sugar	½ tsp. *each* cinnamon and
½ cup butter, melted	nutmeg

Sauce:

¾ cup granulated sugar	2 tbsp. cornstarch
1 cup cold water	1 tsp. grated orange rind

Filling:

1½ cups chopped rhubarb	1 cup sliced strawberries
1 cup sliced peeled apples	1 cup blueberries

Crust: In a bowl, combine flour, oats, brown sugar, cinnamon and nutmeg; stir in butter. Press half of the mixture over bottom of greased 9-inch square cake pan.

Sauce: In a small saucepan, combine sugar with cornstarch; whisk in water and orange rind until smooth. Bring to boil, reduce heat to medium-low and cook for 5 minutes or until thickened and clear, whisking constantly.

Filling: In bowl, toss together chopped rhubarb, strawberries, apples and blueberries; arrange over base. Pour sauce over top; sprinkle with remaining flour mixture. Bake in 350°F oven for 50 to 60 minutes or until fruit is tender and topping is golden. Serve warm.

Makes 8 servings.

"Children can stand vast amounts of sternness. It is injustice, inequity and inconsistency that kills them."

BLUEBERRY GRUNT

Joan Gardner, Aldergrove

6 tbsp. butter
1 cup all-purpose flour
1 cup sugar
2 tsp. baking powder
Dash of salt

¾ cup milk
1 tsp. vanilla
4 cups blueberries
Whipping cream

Melt butter in a 12" x 7" baking dish. Stir together flour, sugar, baking powder, and salt. Stir in milk and vanilla. Pour mixture over melted butter in pan. Sprinkle blueberries over the top of cake mixture. Bake in a 375°F oven for 30 minutes or until golden brown. Serve with whipped whipping cream.

Makes 6 - 8 servings.

GOOSEBERRY CRUMBLE

1 tsp. butter
½ cup sugar
2 tbsp. water
3 oz. butter, cut in small pieces

1½ lb. gooseberries, drained
½ cup flour
¼ cup brown sugar

Preheat the oven to moderate (350°F) oven. Lightly grease a medium-sized baking dish with butter. Arrange the gooseberries in the dish and sprinkle them with the sugar and water.

Sift the flour into a medium-sized mixing bowl. With your fingertips, rub the butter into the flour until the mixture resembles fine bread crumbs. Stir in the brown sugar. Cover the gooseberries with the crumble mixture and place the dish in the oven. Bake for 45 minutes, or until the crumble is golden brown. Remove the dish from the oven and serve at once with cream.

"A mother's patience is like a tube of toothpaste—never quite all gone."

BERRY COBBLER

1 pint fresh raspberries
  (2½ cups)
1 pint fresh blueberries or sliced
  strawberries (2½ cups)
2 tbsp. cornstarch
½ cup no-sugar-added
  raspberry pourable fruit
1 cup all-purpose flour

1½ tsp. baking powder
¼ tsp. salt
⅓ cup milk
⅓ cup butter or margarine,
  melted
2 tbsp. thawed frozen
  unsweetened apple juice
  concentrate

Preheat oven to 375°F. Combine berries and cornstarch in medium bowl; toss lightly to coat. Add pourable fruit; mix well. Spoon into 1½ quart or 8-inch square baking dish. Combine flour, baking powder and salt in medium bowl. Add milk, butter and concentrate; mix just until dry ingredients are moistened. Drop six heaping tablespoonfuls of batter evenly over berries; sprinkle with nutmeg. Bake 25 minutes or until topping is golden brown and fruit is bubbly. Cool on wire rack. Serve warm or at room temperature.

BLACKBERRY CRUNCH

Joyce O'Hara, Ireland

4 cups blackberries
6 tbsp. sugar
3 cups toasted, sifted
  bread crumbs

¾ cup icing sugar
½ cup melted butter
1 tsp. cinnamon

Mix blackberries gently with sugar. Blend bread crumbs, icing sugar, butter and cinnamon. Pour berries in baking dish and cover with bread crumb mixture. Bake at 325°F for 30 minutes.

"The problem may be less the path we choose than the way we travel it."

BLUEBERRY RHUBARB BUCKLE

Rita Pollock, Coquitlam.

"This is the simplest and best recipe I have ever found for 'buckle'. It has an almost shortbread texture. I inadvertently found out the wonderful taste of the above combination when I was down to one cup of blueberries and one cup of rhubarb. Of course, with 'buckle' any kind of berry can be used but I like this best."

Cake:
1 egg, beaten
½ cup sugar
½ cup shortening
2 cups flour
¼ tsp. salt
½ cup milk
2½ tsp. baking powder

Topping:
1 cup sugar
1 cup flour
½ cup butter
1 tsp. cinnamon (optional)

1 cup blueberries
1 cup rhubarb

Mix together egg, ½ cup sugar and shortening. Blend in flour, salt, milk and baking powder. Press the mixture into a 9x13-inch ungreased pan. Cover with blueberries and rhubarb. Use more fruit if desired. For topping, combine sugar, flour, butter and cinnamon. Mix until crumbly and spread evenly over berry mixture. Bake at 350°F for 1 hour. Enjoy as is or serve with cream, ice cream or whipping cream.

"The instruction received at the mother's knee is never effaced entirely from the soul."

BLUEBERRY COBBLER

Evelyn Harris, Vancouver.

1 cup flour	1 egg, slightly beaten
2 tbsp. sugar	¼ cup milk
1½ tsp. baking powder	½ cup sugar
¼ tsp. salt	2 tbsp. cornstarch
¼ tsp. lemon peel, grated	1½ cups water
¼ cup butter	4 cups blueberries, drained

Sift together flour and the 2 tbsp. of sugar, baking powder, salt and lemon peel in large bowl. Cut in butter with a pastry blender until crumbly. Mix eggs and milk; add all at once to dry ingredients. Stir just until dry ingredients are moistened. Combine ½ cup sugar, cornstarch and water in a small saucepan. Cook until thick and bubbly. Add blueberries; cook over medium heat for 5 minutes. Pour into a 2 quart casserole dish. Drop topping in 8 mounds on top of hot fruit. Bake uncovered from 25 minutes at 425°F until light brown.

Makes 8 servings.

BLUEBERRY BUCKLE CAKE

¾ cups sugar	½ tsp. ground nutmeg
¼ cup vegetable shortening	1½ cups blueberries,
2 eggs	fresh or frozen, drained
1½ cups milk	
1½ cups all-purpose flour	**Crumb mixture:**
2 tsp. baking powder	¼ cup butter, softened
½ tsp. salt	⅓ cup flour
¼ tsp. ground cloves	½ tsp. ground cinnamon

continued next page

(Blueberry Buckle Cake) …Mix sugar, shortening, eggs and milk until well blended. Stir in flour, baking powder, salt, cloves and nutmeg. Fold in blueberries. Spread batter into a greased 9-inch square pan. Combine remaining ingredients and mix until crumbly. Sprinkle crumbs over batter. Bake in a preheated moderate oven (375°F) for 45 to 50 minutes or until top springs back when lightly touched. Serve warm, cut into squares. If desired, serve with lemon sauce.

See sauce section for Lemon Sauce.

GOOSEBERRY CHARLOTTE

Joyce O'Hara, Ireland

2 cups fine stale bread crumbs
¼ cup butter
1 lb. gooseberries
2 tbsp. water
1 tbsp. lemon juice
½ cup brown sugar

Melt butter and mix in with the bread crumbs; set aside. Stew gooseberries gently in the water until tender. Put a layer of buttered crumbs in a greased pie dish, then a layer of gooseberries. Sprinkle with sugar and lemon juice. Continue layering until dish is full, finishing with a layer of crumbs on top. Cover and bake for 30 minutes at 375°F, then uncover and continue baking until crisp on top.

BLACKBERRY BETTY

3 cups blackberries
¼ cup sugar
1 tsp. orange rind, grated
2 tbsp. water
Slices of stale sponge cake
2 eggs, beaten
3 cups milk

continued next page

(Blackberry Betty) ...Stew blackberries gently with the sugar, orange rind and water until pulpy. In a well greased deep-dish pie plate, layer thin slices of cake on bottom and top with fruit mixture. Continue layering until all cake and berries are used, ending with cake layer. Combine eggs and milk well. Pour over cake layers. Place pie plate in a larger pan in case of boil over. Bake at 350°F for 45 to 50 minutes until cake has set and top is browned.

BLUEBERRY BETTY

2 cups fresh blueberries	4 cups white bread, cubed
Juice of 1 lemon	¼ cup granulated sugar
½ cup brown sugar, packed	1 tsp. ground cinnamon

Mix berries, lemon, and brown sugar. Spread half in shallow baking dish. Mix bread cubes, sugar, and cinnamon and put half over berries. Sprinkle with rest of berries and top with rest of bread cubes mixture. Bake in preheated moderate oven (350°F) for 25 to 30 minutes. Serve warm with cream.

Serves 4 - 6.

STRAWBERRY BETTY

4 cups strawberries	4 cups white bread, cubed
Juice of ½ a lemon	¼ cup granulated sugar
²/₃ cup brown sugar, packed	1 tsp. lemon rind, grated
2 tbsp. butter	

Wash and hull berries. Mix berries with lemon juice and brown sugar. Put in shallow 1½ quart baking dish. Mix bread cubes, granulated sugar, and grated lemon rind. Sprinkle over strawberries and dot with butter. Bake in preheated moderate oven (350°F) for 25 to 30 minutes. Serve warm, with cream.

LOW-CAL BERRY TRIFLE

Custard:

²/₃ cup low-calorie
sweetener
¼ tsp. salt
2 eggs, lightly beaten
1 tsp. grated orange rind

¼ cup cornstarch or minute
tapioca
1¾ cups skim milk
1 tsp. vanilla

Trifle:

½ pkg. low fat pound cake
(about 3 cups cubed)
1 tbsp. orange liqueur or
orange juice

2 tbsp. no sugar added
strawberry jam
3 cups assorted fresh soft
fruit, i.e. strawberries,
raspberries, blueberries, etc.

Custard: Mix sweetener, starch, and salt in heavy saucepan. Stir in milk. Stir over medium heat until boiling. Simmer, stirring, over low heat 2 minutes. Gradually stir into eggs. Return to saucepan; cook and stir 1 minute. Remove from heat, stir in vanilla and orange rind. Chill.

Assemble: Cut pound cake in half lengthwise. Spread with jam. Put layers together; cut into ½ inch cubes. Place in glass bowl (about 2 quart size). Drizzle with liqueur or juice. Arrange fruit on top. Spread custard over fruit. Refrigerate for at least 1 hour.

Makes 8 to 10 servings.

"Unless we think of others and do something for them, we miss one of the greatest sources of happiness."

79

GOOSEBERRY DUMPLINGS

2 cups gooseberries
1 ¼ cups sugar
2 cups water
½ tsp. grated lemon rind

Dumpling Batter

Cream (optional)

Combine berries, sugar, water, and lemon rind in deep medium skillet. Cook slowly for 5 minutes. Drip dumpling batter by tablespoonfuls to make 6 dumplings on top of fruit. Cover and cook gently for 20 minutes. Serve with cream topping. Makes 6 servings.

Dumpling Batter:

1 cup sifted cake flour
2 tbsp. sugar
1 tsp. baking powder

¼ tsp. salt
1 tsp. melted butter
¼ cup milk

Sift dry ingredients together. Add butter and milk. Mix only until flour is moistened.

OLD FASHIONED RASPBERRY BREAD PUDDING

B.C. Raspberry Growers Association

10 slices bread, crustless
2 tbsp. butter
½ tsp. cinnamon
3 tbsp. sugar
1 ½ cups raspberries,
    fresh or frozen with juice

3 large eggs
13 oz. evaporated milk
2 tbsp. sugar
Whipped cream

Set oven at 350°F. In a greased 8 x 10-inch baking dish, butter one side of bread, with buttered side up, place 5 slices in bottom.

continued next page

80

(*Old Fashioned Raspberry Bread Pudding*) ...Sprinkle with cinnamon and sugar and ½ the raspberries and juice. Add remaining bread and remaining raspberries. Beat the egg, milk and sugar in a small bowl. Pour over bread. Bake 35 minutes or until done. Serve with whipped cream.

SUMMER PUDDING

Celia Hill, Surrey.

1½ lbs. soft summer fruits (use mostly red currants, then add raspberries, strawberries, black currants, rhubarb and blackberries)

7 to 8 slices of stale white mixed bread
½ to ⅔ cup sugar (to taste)

Line bottom and sides of 1.5 L soufflé dish with slices of bread, cutting to fill any gaps. Bring fruit and sugar over medium heat to boil (stirring to melt sugar), turn heat down and simmer for 2 to 3 minutes just until juices begin to run. Reserve a few tablespoons of the juice. Pour fruit and juice into dish and place remaining slices of bread over top. Set a plate on top of bread with a weight on it and put it in refrigerator until chilled. Pudding can be loosened and turned out onto a plate for serving. Use juice to soak any bits of bread which are still white. Serve chilled with crème fraîche or whipped cream.

BERRY PUDDING

Evelyn Underwood

1 lb. fresh berries
½ cup butter
¾ cup flour

⅛ to ¼ cup sugar
2 eggs
1 tsp. baking powder

continued next page

(Berry Pudding) …Cook berries a few minutes in juice and with a few tbsp. of water to create a thin sauce. Pour fruit into a 9 x 13-inch baking pan. Cream butter and sugar until soft, add beaten eggs and mix well. Fold in flour and baking powder. Mix to soft consistency and spread over fruit. Bake in oven at 375°F for 40 minutes.

STEAMED BLUEBERRY PUDDING

3 tbsp. butter	1 cup all-purpose flour
¼ cup sugar	1 tsp. baking powder
1 egg	¼ tsp. salt
⅓ cup milk	1 cup fresh blueberries

Cream butter and sugar. Add egg; beat well. Add sifted dry ingredients alternately with milk. Fold in blueberries. Fill greased 6 oz. custard cups or other individual molds ⅔ full. Cover tops with wax paper; tie on firmly with cord. Place on rack in steamer or large kettle containing 1" of boiling water. Cover kettle and steam for 1 hour, adding more boiling water if necessary. Unmold and serve with Foamy Blueberry Sauce. See sauce section.

Makes 4 servings.

STRAWBERRY RICE PUDDING

2 cups milk	¼ cup water
½ cup uncooked rice	3 egg yolks
½ tsp. salt	½ cup sugar
¼ cup rum	½ tsp. vanilla extract
2 cup strawberries, sliced	1 cup heavy cream, whipped
1 envelope unflavored gelatin	2 cups strawberries, halved

continued next page

(Strawberry Rice Pudding) ...Scald 1 ¹/₃ cups of the milk in top part of double boiler. Add rice and salt and cook, covered, over boiling water until rice is soft. Depending on the quality of the rice used, this may take up to 1 hour. Stir rice occasionally. If rice dries out before getting soft, add more milk. Pour rum over sliced strawberries and chill.

Soften gelatin in water. Beat egg yolks with sugar and add remaining ²/₃ cup milk. Cook over simmering, not boiling, water until mixture coats spoon. Stir constantly. Add gelatin and blend until gelatin is completely dissolved. Add gelatin-custard mixture to rice and mix thoroughly. Add vanilla. Chill until mixture is slightly thickened. Fold in sliced strawberries and whipped cream. Chill. Pile into sherbet glasses; garnish with berry halves.

RASPBERRY PUDDING

2 cups raspberry juice
1 cinnamon stick
¼ cup cornstarch
Cream

¼ cup sugar, to taste
¼ tsp. salt
Red food coloring (optional)

Bring juice, sugar, cinnamon stick, and salt to a boil. Mix the cornstarch in a small amount of cold water and stir into the hot juice. Bring mixture to a quick boil and remove from heat as soon as it thickens. Add a few drops of red food coloring if necessary. Pour into pudding dish or individual glasses. Sprinkle the top with additional sugar to keep a skin from forming. Chill. Serve with cream. Note: Sugar according to taste, the pudding should be tart rather than sweet.

"When we cannot find contentment in ourselves, it is useless to find it in others."

FROZEN BLUEBERRY PARFAIT

½ cups blueberries, fresh
  or frozen, drained
¼ cup sugar
2 tbsp. cornstarch
¼ cup water

1 tbsp. lemon juice
8 oz. plain yogurt
3 peaches, coarsely
  chopped

In a small saucepan, stir together berries and sugar, Stir cornstarch and water until smooth. Add to fruit mixture. Cook over medium heat, stirring constantly, until mixture comes to a boil. Boil 1 minute. Stir in lemon juice. Cool. Fold yogurt into cooled fruit. In parfait glasses, alternately layer berry mixture with peaches. Freeze. Remove from freezer ½ hour before serving. Serves 4.

BERRY MEDLEY

Fresh or frozen or use your own combo...

1 cup raspberries
1 cup blueberries
1 cup strawberries, halved
1 cup black currants

¹/₃ cup sugar
½ vanilla bean
1 tbsp. lemon juice
1 tbsp. water

In a large saucepan, combine all ingredients. Cook over low heat gently stirring until sugar dissolves. Cool. Discard vanilla bean. Pour into a glass serving bowl or parfait glasses, cover and refrigerate. Serve with wafer cookies, if desired.

BLUEBERRIES ROMANOFF

2 cups fresh or frozen
  blueberries, rinsed
  and drained
1 pint vanilla ice cream, softened

¹/₃ cup sugar
1 cup heavy cream, whipped
¼ cup port wine

continued next page

84

(Blueberries Romanoff) …Pour blueberries into a bowl and sprinkle with sugar. Let stand 10 minutes. In another bowl mix whipped cream with softened ice cream. Stir in port wine. Drain blueberries and fold into cream mixture. Place bowl into freezer and freeze for 1 hour. Spoon mixture into parfait glasses and serve garnished with additional blueberries. Can also be spooned over squares of yellow cake or into cantaloupe halves. Serve at once. Serves 8.

FROZEN GOOSEBERRY FOOL

1 lb. fresh gooseberries
1 cup water
Sprig of mint
A little green food coloring

½ cup sugar
1 cup double cream,
   lightly whipped

Wash and stem gooseberries. Combine berries with water and mint. Cover and simmer gently for 15 minutes or until soft. Remove from heat. Stir in sugar and a little green coloring. Discard mint sprig. Purée in a blender, sieve to remove all pips. Cool. Blend mixture with whipped cream. Place in small serving dishes, refrigerate until ready to serve. Garnish with sprig of mint or extra gooseberries. Freezes well. Makes 4 servings.

GINGERED FRESH
BLUEBERRY COMPOTE

2 cups fresh blueberries
1 cup fresh orange juice
1 tbsp. fresh lemon juice

¼ cup icing sugar
2 tbsp. ginger root, minced
Fresh mint leaves, for garnish

Wash and drain blueberries and place in a serving bowl. Combine juices, sugar and ginger root and pour over blueberries. Chill for 1 to 2 hours. Serve in sherbet glasses or fruit dishes. Garnish with fresh mint leaves. Makes 6 servings.

STRAWBERRY AND TOFU FOOL

½ lb. fresh strawberries, hulled   ½ lb. silken tofu
2 tbsp. clear honey   2 tsp. lemon juice

Reserve three to four strawberries for decoration and place the remainder in a blender or food processor. Blend very briefly, then pour into a bowl.

Blend the tofu, honey and lemon juice until smooth. Pour onto the strawberries and stir to mix. Pour into dessert glasses and chill. Just before serving, place a whole strawberry on top of each. Substitute other soft fruits, such as black currants, red currants, or raspberries.

GOOSEBERRY FOOL

Evelyn Underwood

1½ lbs. gooseberries   6 oz. sugar
3 tbsp. water   1 cup thick custard

Top and tail the berries. Cook in water. When nearly tender, add sugar and cook until tender. Pass berries through sieve and leave aside until cooled. Whip custard slightly and fold into gooseberry purée. Pour into dessert glasses and serve cold.

"One of the purest and most enduring of human pleasures is to be found in the possession of a good name among one's neighbors and acquaintances."

Sauces and Toppings

STRAWBERRY HARD SAUCE

²/₃ cup butter
1 cup sliced strawberries

2 cups sifted confectioners'
sugar

Cream butter and gradually beat in sugar. When thoroughly blended, stir in sliced strawberries.

Serves 8.

BLUEBERRY SAUCE

Serve on pancakes, waffles, blintzes, ice cream or pudding.

2 cups blueberries
2 tbsp. sugar
1 tbsp. corn starch
¹/₈ tsp. ground nutmeg

¼ cup orange juice
¼ cup water
¼ tsp. grated orange peel
Pinch of salt

Combine all ingredients in saucepan. Cook and stir over medium heat 4 to 5 minutes or until thickened.

Makes 2 cups.

STRAWBERRY VELVET SAUCE

1¼ cups strawberries, sliced
frozen with juice

½ cup heavy cream

Put thawed berries in electric blender and whirl for a few seconds. On low speed, gradually add cream, whirling until thick and smooth.

"All that is necessary to break the spell of inertia and frustration is to 'act as if it were impossible to fail'."

FOAMY BLUEBERRY SAUCE

¼ cup butter
1 cup sugar
1 cup blueberries, crushed

1 tbsp. lemon juice
¼ tsp. salt
1 egg white

Cream butter and sugar. Add blueberries and lemon juice and beat well. Cook in top part of double boiler over boiling water for 5 minutes. Add salt to egg white; beat until stiff. Fold into warm berries and serve at once on pudding. Makes 2 cups.

RASPBERRY SAUCE

4 cups raspberries
½ cup sugar
Lemon juice

Water
2 tbsp. cornstarch

Mash and sieve berries; add water to make 2 cups. Add sugar and cornstarch, mixed. Cook until thickened, stirring constantly. Add lemon juice. Cool. Chill.

LOW-CAL RASPBERRY SPREAD

South Alder Farms Ltd., Aldergrove.
This is great for cake filling or over ice cream.

1½ tsp. unflavored gelatin
Artificial sweetener
  (equivalent to 8 to 12 tsp. sugar)

¼ cup water
2 cups fresh raspberries sliced

Sprinkle gelatin over water to soften. Let stand 5 minutes. Place berries in a medium saucepan; bring to boil. Cook, stirring occasionally for 5 minutes. Add softened gelatin and sweetener. Stir until gelatin dissolves. Skim off foam. Ladle into hot sterilized jars leaving ½ inch headspace. Bottle, seal and store in refrigerator or freezer.

LEMON SAUCE

(for Fresh Blueberry Cake)

½ cup sugar
1 tbsp. cornstarch
¼ tsp. salt
¼ cup cold water
¾ cup boiling water

3 tbsp. fresh lemon juice
1 tsp. lemon rind, grated
½ tsp. vanilla extract
2 tbsp. butter

Combine sugar, cornstarch, and salt. Gradually stir in cold water. Gradually stir in boiling water and cook for 3 minutes, or until smooth, clear and slightly thickened. Stir in remaining ingredients. Serve over Fresh Blueberry Cake.

Makes 1½ cups.

ANNA'S BERRY TOPPING

250g pkg. cream cheese, softened
Demerara Sugar, to taste

Whip cream cheese and sugar together. Serve over fresh berries as a quick dessert or use as a dip with fresh strawberries.

"If you spend your life trying to be the person everyone likes, you waste the person you are."

89

Pies and Tarts

CHERRY & BLUEBERRY PIE

1 - 15 oz. pkg. Ready Pie Crusts
1 tsp. flour

Topping:
1 tsp. milk
1 tbsp. sugar
1/8 tsp. cinnamon
1 tbsp. butter

Filling:
¾ cup sugar
1½ cups blueberries
¼ cup cornstarch
2-16 oz. cans pitted pie cherries, drained, reserving 1 cup liquid.

Prepare pie crust according to package directions for 2-crust pie using 9-inch pie pan. Heat oven to 400°F.

In medium saucepan, combine sugar and cornstarch. Gradually add reserved cherry liquid. Cook over medium heat until mixture just begins to boil, stirring occasionally. Boil 1 minute until thick and clear, stirring constantly. Remove from heat. Stir in butter. Gently stir in cherries. Place blueberries (fresh or frozen, thawed and drained) in bottom of pie crust-lined pan. Top with cherry mixture.

Using 1" special occasion shaped cutter, cut 4 shapes from second crust. Carefully place crust on top of filling. Flute. Decorate with shape cutouts. Brush crust and cutouts with milk. Combine sugar and cinnamon, sprinkle over crust and cutouts.

Bake at 400°F for 40 - 55 minutes or until golden brown. Cover edge of pie crust with strip of foil during last 10 - 15 minutes to prevent excessive browning is necessary. Cool well before serving.

Serves 8

"A little boy becomes a man a lot quicker than his parents think, but not nearly as quickly as he thinks."

BLUEBERRY PIE ONE

Elizabeth Kuntz

¾ cup sugar
1 tsp. ground cinnamon
1 tsp. grated orange peel
2 tbsp. orange juice
butter

¼ cup flour
4 cups fresh or frozen
  blueberries, thawed
2 *each* 9-inch pie pastry, 1 tbsp.
  unbaked

Combine sugar, flour cinnamon and orange peel. Lightly toss with blueberries. Place in pastry lined pie plate. Sprinkle with orange juice, dot top with butter. Roll out remaining pastry. Cut into ½ inch strips. Arrange in lattice pattern on top of pie. Moisten edge of lower crust. Fold over lattice strips, seal and flute.

Bake at 425°F for 10 minutes. Lower to 350°F for 35 - 40 minutes or until golden brown and filling begins to bubble.

Serves 8

RASPBERRY PEACH PIE

1 cup water
¾ cup sugar
2 tbsp. cornstarch
3 tbsp. peach-flavored gelatin
9-inch butter or graham
  cracker crust

1 cup fresh raspberries
3 cups fresh peaches,
  peeled, sliced
Whipped cream or
  frozen whipped topping

In medium saucepan, combine water, sugar and cornstarch. Cook over medium heat until mixture boils and thickens. Remove from heat. Stir in gelatin until it dissolves. Cool slightly. Fold in raspberries and peaches. Refrigerate 20 minutes. Pour into crust-lined pan. Refrigerate 2 hours or until set. Serve with whipped cream. Store in refrigerator.

Serves 8

MICROWAVE BLUEBERRY PIE

Jocelyn Methot

1 9-inch pie crust, baked

Filling:

3 cups fresh blueberries	½ cup sugar
½ tsp. lemon zest	¼ cup all-purpose flour
¼ tsp. cinnamon	¼ tsp. nutmeg (or less)

Topping:

½ cup quick-cooking oats	2 tbsp. butter
2 tbsp. brown sugar	½ tsp. cinnamon

Mix together filling ingredients and spread evenly in pie crust. Blend together topping ingredients and sprinkle on top of fruit mixture. Cook for 5 minutes on 'High", then for 4 minutes on 'Medium', or until the mixture boils and thickens. For variety, use 4 cups fresh strawberries instead of blueberries, and increase sugar to ⅔ cup.

GOOSEBERRY PIE

3 cups fresh gooseberries	1½ cups sugar
3 tbsp. quick-cooking tapioca	¼ tsp. salt
2 *each* 9-inch pastry, unbaked	2 tbsp. butter

Stem and wash gooseberries. Crush ½ cup. Mix crushed berries with sugar, tapioca and salt. Bring to a bubbly boil, stirring constantly. Cook 2 minutes. Add remaining whole berries. Prepare pastry for 2-crust 9-inch pie. Line pie plate with pastry and fill. Dot with butter. Adjust top crust, cut slits, seal.

Bake at 400°F for 35 minutes.

"A young child, a fresh uncluttered mind, a world before him—to what treasures will you lead him?"

BLUEBERRY PIE TWO

1 cup sugar
¼ cup all-purpose flour
¼ tsp. salt
½ tsp. cinnamon
½ tsp. grated lemon rind

4 cups fresh blueberries
1 tbsp. lemon juice
2 *each* 9-inch pastry, unbaked
2 tbsp. butter

Prepare pastry and line 1-9-inch pie plate with pastry rolled ⅛-inch thick. Combine all other ingredients except butter and turn into pie plate. Dot with butter. Cover with remaining pastry. Trim, turn under and flute edge. Cut a gash in top of crust to allow for escape of steam. Bake in preheated oven (425°F) for 40 minutes.

DEEP-DISH MINI-PIES

1 recipe pastry dough, unbaked
1 egg yolk, slightly beaten
2 tbsp. quick-cooking tapioca
2 tbsp. sugar

Dash of lemon juice
⅛ tsp. salt
4 cups berries
in heavy syrup

Roll out pastry and cut 4 rounds to fit tops of serving dishes. Put on ungreased baking sheet, prick with fork and brush with beaten egg yolk and a little cold water. Bake at 425°F for 10 minutes.

Mix tapioca, sugar, lemon juice, salt and berries. Cook, stirring constantly until thickened. Pour into 4 serving dishes and top with precooked pastry tops.

Makes 4 servings.

"The only thing children wear out faster than shoes are parents and teachers."

MILE HIGH RASPBERRY PIE

B.C. Raspberry Growers Association

¾ cup sugar
2 egg whites
2 cups frozen raspberries
1 tsp. vanilla
1 tbsp. lemon juice

9-inch pastry or crumb shell,
 baked
1 cup whipping cream
Pinch of salt

Combine sugar, egg whites, raspberries (partially thawed), vanilla, lemon juice and salt in a large mixer bowl. Beat at high speed for 15 minutes, until thick and voluminous. Fold in whipped cream and pile mixture in baked pie shell. Freeze several hours. After pie is frozen, wrap in foil. To serve, remove from freezer, slice and serve immediately. Will keep a few days in freezer.

Makes 6 - 8 servings.

PEACH BLUEBERRY PIE

Sue Thumma

1 cup sugar
¹/₃ cup all-purpose flour
½ tsp. ground cinnamon
¹/₈ tsp. ground allspice
3 cups peaches,
 peeled & sliced

1 tbsp. butter
2 *each* 9-inch pastry, unbaked
1 cup fresh blueberries
Milk
Cinnamon-sugar

In a bowl, combine sugar, flour, cinnamon and allspice. Add the peaches and blueberries; toss gently. Line pie plate with bottom crust; add the filling. Dot with butter. Top with a lattice crust. Brush crust with milk; sprinkle with cinnamon-sugar. Bake at 400°F for 40 to 45 minutes or until crust is golden brown and filling is bubbly. Cool completely before serving.

Makes 6 - 8 servings.

STRAWBERRY CHOCOLATE PIE

Lauretha Rowe

3 oz. semi-sweet chocolate
1 tbsp. butter
6 oz. cream cheese, softened
½ cup sour cream
9-inch pastry shell, baked

3 tbsp. sugar
½ tsp. vanilla extract
3 - 4 cups fresh strawberries
½ cup strawberry jam

In a saucepan, melt 2 oz. chocolate and butter over low heat. Stir constantly. Spread or brush over the bottom and up the sides of pastry shell. Chill. In a mixing bowl, beat cream cheese, sour cream, sugar and vanilla until smooth. Spread over chocolate layer. Cover and chill for 2 hours. Arrange strawberries, tip end up, atop the filling. Brush jam over strawberries. Melt the remaining chocolate and drizzle over all. Makes 6 - 8 servings.

STRAWBERRY SUPREME PIE

Elsie Driediger, Kamloops.

Crumb Shell:

1½ cups graham wafer crumbs
¼ cup sugar

½ cup melted butter

Filling:

1 envelope gelatin
¼ cup cold water
½ cup hot water
2 egg whites

¾ plus ½ cup sugar
¼ tsp. salt
1 cup mashed strawberries
1 cup whipped cream

Crumb Shell: Mix crumbs, melted butter and sugar. Press into 9-inch pie plate and either bake 8 minutes at 350°F or chill thoroughly. Beat egg whites. Gradually add ½ cup sugar until stiff peaks form.

continued next page

(Strawberry Supreme Pie) ...**Filling:** Soften gelatin in cold water and then dissolve in hot water. Add ¾ cup sugar, strawberries. Add salt. Cool until mixture begins to thicken, then fold in whipped cream and egg whites. Pour into shell. Chill and serve.

STRAWBERRY GLACÉ PIE

1 cup biscuit mix
¼ cup butter
3 tbsp. boiling water
4 cups strawberries
1 cup water, divided
1 cup sugar

Garnish:
3 oz. cream cheese, softened
2 tbsp. heavy cream
¼ tsp. vanilla extract

9-inch pastry shell, unbaked

Put biscuit mix and butter in bowl. Add boiling water and stir vigorously with fork to form a ball. With fingers and heel of hand, pat evenly into 9-inch pie pan. Bring dough up over edge of pan and flute edge. Bake in preheated very hot oven (450°F) for 8 to 10 minutes; cool.

Wash, drain and hull berries. Simmer 1 cup of the berries and ²/₃ cup of the water for 3 minutes, or until berries soften. Blend sugar, cornstarch, and remaining water. Stir into berry mixture and cook. Stir constantly until thickened. Put remaining berries in pie shell. Cover with cooked mixture and chill for 2 hours, or until firm.

Garnish: Beat cream cheese until fluffy. Add cream and vanilla. Force through pastry tube to garnish pie edge.

Makes 6 - 8 servings.

"Once you're a grown-up, no one can kidnap your soul unless you act as an accomplice."

STRAWBERRY SNOWY PIE

4 cups strawberries
1¼ cups sugar
½ cup water
2 egg whites

9-inch pastry, baked
Pinch of salt
½ tsp. cream of tartar
½ tsp. vanilla

Wash and hull strawberries; arrange in pastry shell, putting prettiest berries in center. Mix sugar, water and cream of tartar in saucepan. Cover and bring to boil. Uncover and cook until syrup spins long threads (240°F on a candy thermometer). Add salt to egg whites and beat until stiff. Gradually pour syrup onto whites. Beat constantly until mixture forms stiff peaks. Add flavoring and pile on pie, leaving center uncovered. Cool, but do not refrigerate.

Makes 6 - 8 servings.

STRAWBERRY CREAM PIE

9-inch pastry shell, baked
½ cup blanched almonds,
    slivered and toasted
Cream filling, 1 recipe
2 cups strawberries, halved

½ cup strawberries, crushed
½ cup water
¼ cup sugar
2 tsp. cornstarch
Red food coloring

Cover bottom of cooled pastry shell with almonds. Fill with chilled Cream filling. Arrange 2 cups of halved strawberries on filling.

Glaze: Combine ½ cup crushed strawberries and water. Cook 2 minutes and put through a sieve. Mix sugar and cornstarch. Gradually stir in berry juice. Cook, and stir until thick and clear. Tint to desired color with food coloring. Cool slightly. Pour over halved strawberries. Keep refrigerated until serving time.

Makes 6 - 8 servings.

CREAM FILLING

for Strawberry Cream Pie

½ cup sugar
3 tbsp. all-purpose flour
3 tbsp. cornstarch
½ tsp. salt'

2 cups milk
1 egg, beaten slightly
½ cup whipped cream
1 tsp. vanilla extract

Mix sugar, flour, cornstarch and salt. Gradually stir in milk. Stirring constantly, bring to a boil. Reduce heat and cook and stir until thick. Stir a little of hot liquid into egg; return egg mixture to hot liquid. Bring just to boiling, stirring constantly. Cool, then chill. Beat well. Fold in whipped cream and vanilla.

FRESH STRAWBERRY PIE

Rita Reimer

3 cups strawberries, whole
9-inch pastry shell, baked
1 cup strawberries, crushed
¾ cup water

3 tbsp. cornstarch
1 cup sugar
⅛ tsp. salt
1 tsp. lemon juice

Arrange whole strawberries in pastry shell. In saucepan, combine crushed strawberries and water. Simmer 3 to 4 minutes. Strain through sieve to remove seeds. Add water to strained berry juice to yield 1 cup. Combine cornstarch, sugar, salt, and strained berry juice in saucepan. Bring to a boil and cook until thick and clear. Remove from heat and cool. Add lemon juice. Spoon glacé over berries. Let cool. Garnish with whipping cream.

"Obsessing is a poor substitute for action."

TRIPLE BERRY CRISP PIE

Maxine L. Parker, Langley.

Filling:

1 cup strawberries
2 cups raspberries
3 cups blueberries

1 cup sugar
3 tbsp. cornstarch
¼ tsp. nutmeg

Topping:

¾ cup flour
½ cup quick-cooking oats

1 cup brown sugar
½ cup cold butter

Crust:

1-crust 9-inch deep-dish crust, baked and cooled

Mix together filling ingredients and pour into pre-baked crust. Blend topping ingredients and rub together to form crumbs. Spread over berries. Place pie plate on cookie sheet to catch boil-over. Bake 40 minutes at 375°F.

BLASPBERRY PIE

Marge Bekker, Grand Forks.

¾ to 1 cup sugar
4 cups raspberries or
   blueberries or mixed
2 tbsp. butter or margarine

3 to 4 tbsp. flour
¹/₈ tsp. salt
1 tbsp. lemon juice
2 *each* 9-inch pastry, unbaked

In bowl, combine sugar, flour and salt. Spread half of mixture over pastry in pie plate. Top with berries; sprinkle remaining sugar mixture over top of berries; sprinkle lemon juice over all. Dot with butter. Roll, fit and seal upper crust. Bake at 425°F for 30 to 40 minutes or until done. Serve with vanilla ice cream or sweetened whipped cream. Note: if using blueberries, 1 tsp. cinnamon may be added to sugar mixture.

STRAWBARB PIE

Darlene Pekrul, Langley

2 *each* 9-inch pastry shells,
   unbaked
3 beaten eggs
1¼ cups sugar
¼ cup all-purpose flour
¼ tsp. salt

2½ cups rhubarb, sliced
1½ cups fresh strawberries
1 tbsp. butter
½ tsp. nutmeg

Combine eggs, sugar, flour, salt, and nutmeg. Mix well. Combine rhubarb and strawberries. Line pie plate with pastry and fill with fruit. Pour egg mixture over. Dot with butter. Top with lattice crust, crimping edges high. Bake in 400°F oven about 40 minutes. Fill openings in lattice crust with whole strawberries. Serve warm.

VERY RASPBERRY PIE

Kathy Jones

Raspberry Topping:
6 cups fresh raspberries, divided 1 cup sugar
3 tbsp. cornstarch   ½ cup water

Cream Filling:
8 oz. cream cheese, softened   1 cup whipped topping
1 cup icing sugar
1 9-inch graham cracker crust   Fresh mint, optional

Topping: Mash about 2 cups raspberries to measure 1 cup; place in saucepan. Add sugar, cornstarch and water. Bring to a boil, stirring constantly. Cook and stir 2 minutes longer. Strain to remove berry seeds if desired. Cool to room temperature, about 20 minutes.

continued next page

100

(Very Raspberry Pie) ... **Filling:** Beat cream cheese, whipped topping and icing sugar in a mixing bowl. Spread in bottom of crust. Top with remaining raspberries. Pour cooled raspberry sauce over top. Refrigerate until set, about 3 hours. Store in the refrigerator. Garnish with mint if desired. Makes 6 - 8 servings.

LATTICE-TOPPED BLUEBERRY PIE

¾ cup sugar
¼ cup flour
1 tsp. ground cinnamon
4 cups fresh blueberries
1 tsp. orange peel, grated

2 *each* 9-inch pastry shells, unbaked
2 tbsp. orange juice
1 tbsp. butter

Combine sugar, flour, cinnamon and orange peel. Lightly toss with blueberries. Place in pastry-lined pie plate. Sprinkle with orange juice and dot top with butter. Roll out remaining pastry and cut into ½-inch strips. Arrange in lattice pattern on top of pie. Moisten edge of lower crust and fold over lattice ends. Seal and flute edges. Bake at 425°F for 10 minutes. Lower heat to 350°F and bake 35 to 40 minutes or until crust is golden brown and filling begins to bubble.

Makes 6 -8 servings.

CURRANT PIE

4 cups fresh currants
1 cup sugar
3 tbsp. quick-cooking tapioca
¼ cup water

⅛ tsp. salt
8-inch pastry shell, baked
3 egg whites
¼ tsp. salt
6 tbsp. sugar

continued next page

(Currant Pie) …Wash and stem currants. Combine currants, 1 cup sugar, tapioca, water and salt. Cook for 15 minutes, or until tapioca is transparent. Cool. Pour into pie shell. Beat egg whites and salt until foamy. Add 6 tbsp. sugar gradually, 1 tbsp. at a time, continuing to beat until stiff and glossy. Spread over pie, making sure meringue is spread out to the crust. Bake in preheated slow oven (300°F) for 25 minutes, or until meringue is firm.

Makes 6-8 servings.

RASPBERRY CREAM PIE

2 cups raspberries, crushed
Water
²/₃ cup sugar
3 tbsp. cornstarch
Dash of salt

9-inch pastry shell, baked
4 cups raspberries, whole
Garnish:
1 cup whipped cream
Mint sprigs
Raspberries, whole

Force crushed raspberries through a sieve. Add water to berry juice to make 1½ cups. Mix sugar, cornstarch, and salt. Add to berry juice. Cook, stirring constantly, for 5 minutes, or until thickened. Cool. Arrange 4 cups raspberries in baked pie shell. Pour on cooked cornstarch mixture. Chill. Garnish with whipped cream, whole raspberries and mint sprigs.

Makes 6 - 8 servings.

BLACKBERRY PIE

4 cups fresh blackberries
1½ cups sugar
¹/₈ tsp. salt

1½ tbsp. all-purpose flour
2 each 9-inch pastry, unbaked
1 tbsp. butter

continued next page

102

(Blackberry Pie) ...Mix blackberries with sugar, salt and flour. Fill pastry-lined pie pan. Dot with butter. Adjust top crust. Bake at 450°F for 10 minutes, than at 350°F for 25 minutes.

Makes 6 - 8 servings.

FRESH GLAZED CURRANT TARTS

Wanita Murphy, Aldergrove.

4 cups fresh black/red currants, washed, stems removed and drained
6 - 4" baked tart shells
Water

¾ cup sugar
Dash of salt
2 tbsp. cornstarch
3 oz. cream cheese, cut into 6 cubes

Fill 6 tart shells evenly with 3 cups currants. Cook remaining 1 cup currants in 3 tbsp. water for about 5 minutes and force through sieve. Add enough water to make 1¼ cups. Mix sugar, salt and cornstarch. Gradually stir in sieved berries and cook, stirring constantly, over low heat for 5 minutes or until thickened. Cool. Pour cornstarch mixture over currants in tarts. Garnish each with a cube of cream cheese. Chill. Makes 6 servings.

RASPBERRY TARTS

Mrs. Bea Breeden, Coquitlam.

12 tart shells, unbaked
3 tbsp. butter
½ cup sugar
1 egg well beaten

Pinch of salt
1 tsp. almond extract
4 tbsp. rice flour
Fresh raspberry jam

Place unbaked pastry in tart shells. Mix together butter, sugar, egg, salt, almond extract and flour. Put spoonful of raspberry jam in unbaked tart shells. Spoon batter over jam. Bake at 350°F for 30 minutes.

LATTICE STRAWBERRY TARTS

Wanita Murphy, Aldergrove.

4 cups fresh strawberries
6, 4-inch tart shells, unbaked
  plus pastry for lattice

1 cup sugar
2 tbsp. cornstarch
2 tbsp. butter

Cut pastry to fit tart shells. Wash, drain and hull strawberries. Cut larger ones into halves, leaving the smaller ones whole. Add to pastry-filled tart shells. Combine sugar and cornstarch. Pour over strawberry-filled tart shells. Dot with butter. Arrange strips of pastry, lattice fashion, over tarts.

Bake at 450°F for 10 minutes, then reduce heat to 350°F and bake for 20 minutes or until berries are done and pastry is nicely browned. Makes 6 servings.

BLUEBERRY CHEESETART

Blueberry Growers Assoc.

1½ cups vanilla cookie crumbs
6 tbsp. butter, melted
¹/₈ tsp. ground nutmeg
8 oz. cream cheese, at room
  temperature
½ cup sugar
2 eggs
½ tsp. vanilla

2 cups fresh/frozen
  blueberries
¼ cup sugar
¼ cup water
2 tbsp. cornstarch
Pinch of salt
1 tbsp. lemon juice
½ tsp. lemon peel, grated

Combine cookie crumbs, butter and nutmeg. Press into 9-inch spring-form pan. Beat cream cheese, ½ cup sugar, eggs, vanilla and lemon peel until smooth. Spoon into cookie crust. Bake at 375°F for 15 minutes or until firm. Chill for 15 minutes while you make topping.

continued next page

(Blueberry Cheesetart) ...Combine blueberries, ¼ sugar, water, cornstarch and pinch of salt in saucepan. Cook and stir about 4 minutes or until thickened. Stir in lemon juice. Cool to lukewarm. Spread cooled blueberry topping over cheesecake. Refrigerate several hours. Makes 2 cups.

Serves 10 - 12

"Why suffer needlessly because you're not perfect when you can feel satisfied that you are good enough."

Clafoutis, Flans and Tortes

CLAFOUTIS

Limousin, France is famous for the simplicity and abundance of its country style "cuisine" which is based on a wealth of natural resources; for this province has an unlimited supply of milk and butter.

"Clafoutis" has become a kind of emblem of regional solidarity for the Limousins. It is the type of flan mentioned in the following clafoutis recipes.

BLUEBERRY CLAFOUTIS

Mrs. E. M. Handy, Burnaby.

2 cups fresh blueberries
½ cup blanched almonds
⅓ cup sugar
1½ cup milk
⅔ cup flour

3 eggs
⅛ tsp. salt
¼ tsp. almond extract
½ cup whipping cream
2 tsp. sugar

Grease a 10-inch glass pie plate or shallow round baking dish. Spread berries evenly over pan, reserving a few berries as a garnish. In blender or food processor, blend almonds and sugar until finely ground. Add milk, flour, eggs, salt and extract until smooth (about 1 minute). Pour evenly over fruit.

Bake at 350°F for 50 - 55 minutes until top is golden and toothpick comes out clean. Beat whipping cream with 2 tsp. sugar. Spoon over warm clafoutis and garnish with reserved berries.

Serves 6.

FRUIT MEDLEY CLAFOUTIS

Darcy Newman Edgar, Vancouver.

"Served warm or cold or with a side dish of fresh strawberries or with a little cream poured over it, I've made this variation many times for my book club and it's always great!"

3 cups frozen fruit medley,
thawed
1 cup milk
½ cup heavy cream
⅓ cup or more sugar
or artificial sweetener

⅔ cup flour
2 eggs
1 egg yolk
1½ tsp. vanilla extract
Icing sugar for dusting

Arrange fruit in buttered 10-inch deep glass pie plate. Blend remaining ingredients in blender or food processor. Pour over fruit. Bake at 375°F for 35 - 45 minutes or until it is puffed and golden. Sift icing sugar over top and serve warm. Looks very pretty and reheats well in microwave.

STRAWBERRY AND RHUBARB CLAFOUTIS

20-25 fresh strawberries or
unsweetened frozen, thawed
1 cup frozen rhubarb, thawed
1¼ cup sugar
4 eggs

½ cup milk
½ cup flour
⅔ cup cream (15% MF)
1 tbsp. melted butter
Pinch of salt

Cook the rhubarb a few minutes with ½ cup of sugar. Cool and pour in the bottom of a 10-inch well-buttered glass pie plate or shallow baking dish. Place the strawberries on top of the rhubarb evenly. Blend the remaining ingredients in a blender for about 1 minute. Pour over fruit.

Bake at 350°F for about 1 hour until top is golden and toothpick comes out clean. Serve warm. Makes 6 servings.

FRESH BLUEBERRY FLAN

1 cup all-purpose flour
2 tbsp. icing sugar
½ cup butter
1 pkg. (6 serve size) Jell-O
    vanilla pudding and pie-filling

2 cups milk
2 cups fresh blueberries
½ cup apple jelly
2 tbsp. orange liqueur

Combine flour and sugar. Cut in butter until mixture resembles coarse meal. Press firmly into 9-inch flan pan. Chill 30 minutes. Bake at 425 F for 10 minutes or until golden. Cool.

Prepare pudding with milk according to directions on package. Place plastic wrap on pudding surface. Chill 30 minutes. Whisk until smooth, pour into shell. Chill 10 minutes. Top with berries. Melt jelly and liqueur over low heat. Cool and spoon over fruit. Chill before serving.

Makes 8 servings.

RASPBERRY SWIRL TORTE

1¼ cups graham crumbs
¼ cup butter, melted
85g Jell-O raspberry
    jelly powder
250 regular marshmallows

¼ cup milk
2 cups Cool Whip
2 cups fresh raspberries or
    frozen; thawed and drained

Mix crumbs and butter. Press onto bottom of 9-inch springform pan. Chill 10 minutes. Dissolve jelly powder in 1 cup boiling water. Chill until partially set. Melt marshmallows with milk over low heat. Whisk until smooth. Cool slightly. Gently fold topping onto marshmallow mixture.

continued next page

(Raspberry Swirl Torte) ...Fold berries into slightly thickened jelly. Spoon ½ marshmallow mixture over base, top with ½ the jelly. Repeat marshmallow layer. Dollop remaining jelly; gently swirl with knife. Note: Make sure jelly is set to same degree of firmness as marshmallows mixture before layering. Chill 4 hours or overnight.

Serves 10.

STRAWBERRY MERINGUE TORTE

Louise Hartung/Dorothy Skerritt, Langley.

3 egg whites	½ cup pecans, chopped
½ tsp. baking powder	4 cups fresh strawberries
1 cup sugar	Sweetened whipped cream
10 soda crackers, rolled fine	

Heat oven to 300°F. Butter generously a 9-inch pie pan. Beat egg whites with baking powder until frothy. Gradually beat in sugar until whites are stiff. Fold in cracker crumbs and pecans. Spread in pie pan. Bake 30 minutes. Fill with strawberries and top with whipped cream.

Chill several hours before serving.

"There is no life without struggle, but one seeker can help another and together they can make the journey neither could make alone."

DESSERT PIZZA WITH FRESH FRUIT

B.C. Raspberry Growers Assoc.

Crust:

¼ lb. butter or margarine ⅓ cup confectioner's sugar

1 cup flour

Cheese Layer:

8 oz. cream cheese or Neufchatel cheese

¼ cup sugar ¼ tsp. vanilla

Fruit Topping:

1½ cups red raspberries ½ cup blueberries

1 Kiwi fruit, sliced

Glaze:

2 cups red raspberry purée or crushed raspberries in their juice

½ cup sugar 3 tbsp. cornstarch 2 tsp. lemon juice

Crust: Combine flour and sugar. Cut butter into flour and sugar until of crumb consistency. Pat mixture into 12" pizza pan. Bake at 325°F for 15 to 18 minutes. Cool to room temperature.

Cheese Layer: Whip cheese, sugar and vanilla together. Spread evenly over cooled crust.

Fruit Topping: Wash, drain and stem/peel fruit. Arrange fruit on top of cheese layer.

Glaze: Crush berries, then squeeze them through a fine sieve or colander to make purée. Combine sugar and cornstarch. Add to berry purée or crushed raspberries. Cook over low heat until thickened or microwave 4 minutes—medium high. Add lemon juice and cool to room temperature. Pour over slices just before serving. Makes 12 3-inch slices.

STEWED GOOSEBERRIES

4 cups gooseberries   1½ cups water
1½ cups sugar

Wash berries, and remove blossoms and stems. Discard any soft berries. Bring water and sugar to boil, and simmer for 5 minutes. Add berries and simmer for 10 to 15 minutes, stirring occasionally, until slightly thickened. Serve over ice cream or with whipped cream.

FRIED STRAWBERRIES

Rhonda Driediger, Denver, Colorado.

Fresh Strawberries   Tempura Batter
Brown Sugar   Cinnamon

Wash and dry fresh strawberries, remove stems. Place the strawberries in a tempura batter and fry at high heat according to directions on batter package. Drain on paper towel and roll in brown sugar and cinnamon. Serve with Strawberry Velvet Sauce.

STRAWBERRY SOUP

2 cups strawberries, fresh or   1 cup water
  frozen, no sugar,   1 cup apple juice
  partially thawed   ²/₃ cup sugar
1 cup lowfat strawberry yogurt   ½ tsp. cinnamon
¼ cup lowfat sour cream   ⅛ tsp. cloves
2 tbsp. milk

In large saucepan, combine ¾ cup of water, apple juice, sugar, cinnamon and cloves. Cook over high heat until mixture boils. Cool. In blender, combine strawberries and ¼ water. Blend until smooth. Add berry mixture and yogurt to apple juice mixture. Blend well. Cover, refrigerate until cold.

continued next page

Strawberry Soup) ...**To serve:** Pour soup into chilled individual serving dishes. In small bowl, combine sour cream and milk. Using a spoon or squeeze bottle, place 2 tsp. of this mixture in center of each bowl. Using a toothpick or such, swirl mixture around soup to create a design.

Makes 10 - ½ cup servings.

RASPBERRY DELIGHT

Catherine M. Pullen, Burnaby.

Crust:

1 cup graham wafer crumbs
2 tbsp. butter, melted
Mix well. Pack ²/₃ of crust
in a 8-inch square pan.

Cheese Filling:

4 oz. cream cheese
½ cup icing sugar
¹/₈ tsp. salt

Filling:

1 pkg. Raspberry Jell-O
¼ cup sugar
1 tsp. lemon juice
1¼ cup boiling water

2 cups raspberries,
   fresh or frozen, thawed.
1 cup whipping cream

Mix well together filling ingredients. Cool. Add raspberries. Chill slightly.

Cream together cheese filling. Whip up whipping cream. Fold both together. Spread ½ over crumb crust. Cover with chilled jelly mixture. Cool until set. Spread with remaining cream mixture, top with balance of crumbs. Refrigerate until ready to serve.

Serves 8

"You give but little when you give of your possessions. It is when you give of yourself that you truly give."

BLASPBERRY SQUARES

Cake:
1/3 cup butter
3/4 cup sugar
1 egg
1 tsp. almond extract
2 cups all-purpose flour
2½ tsp. baking powder
¼ tsp. salt

Berry Mixture:
1½ cups fresh berries
1 tsp. lemon rind, grated
¼ cup sugar

2/3 cup milk

Mix all cake ingredients in their order. Prepare berry mixture. Place half of batter in a greased and floured 9-inch square cake pan. Spread berry mixture on this and then spread remaining batter on top. Bake in a moderate oven (350°F) for about 40 minutes or until done. Cut into squares and serve with milk or cream.

RASPBERRY JAMMER BARS

B.C. Raspberry Growers Assoc.

½ cup white sugar
½ cup brown sugar, packed
½ cup peanut butter, crunchy
½ cup shortening
1 egg

1¼ cup self-rising flour
½ cup raspberry preserve
1 cup icing sugar
2 tbsp. peanut butter, crunchy
2 - 3 tbsp. water

Blend sugars, ½ cup peanut butter, shortening and egg. Mix in flour. Refrigerate 1 hour. Heat oven to 375°F. Roll dough to ¼-inch thick on lightly floured board. Cut into rectangles 6 x 1½ inch Make indentations down centers of rectangles with wooden spoon handle. Fill with about 2 tsp. preserves. Place on ungreased baking sheet. Bake 8 to 10 minutes. Cool slightly. Mix icing sugar, 2 tbsp. peanut butter and the water. Drizzle over cooled Jammer Bars.

Makes about 14 bars.

BERRY BOUNTY BOWL

1 package Baker's white
  chocolate
2 to 3 cups mixed berries

Orange rind, grated
Whip topping, light
  or regular, thawed

Press piece of foil inside large bowl. Chill in freezer. Partially melt chocolate over hot water (until $^2/_3$ melted). Remove from heat and stir until smooth.

Drizzle half the chocolate over foil to create a lace pattern. Freeze for 5 minutes. Repeat with remaining chocolate and freeze until set. This chocolate lace bowl can be stored in freezer up to 3 months. To serve, carefully lift foil from chocolate. Gently fold in mixed berries and orange rind into topping. Spoon carefully into lace bowl and top with extra berries.

FROZEN BERRY DELIGHT

Be creative, mix those berries up!

Crust:

2 cups chocolate wafer
  cookie crumbs
$^1/_3$ cup butter, melted
¼ cup sugar

Filling:

1 cup chocolate fudge
  sauce, softened
4 cups vanilla ice cream
  slightly softened
2 cups orange sherbet,
  slightly softened

Topping:

1 cup whipped topping,
  frozen, thawed

12 oz. frozen berries
  without sugar

In medium bowl, combine all crust ingredients. Mix well. Reserve ¼ cup for topping. Press remaining crust in a 13" x 9-inch pan. Refrigerate for 15 minutes.

continued next page

114

(Frozen Berry Delight) ... Spread chocolate fudge sauce over crust. Spoon ice cream over chocolate. Place spoonfuls of sherbet randomly over ice cream; swirl gently into ice cream. Top with berries, pressing gently into sherbet. Spread whipped topping over ice cream sherbet mixture. Top with reserved crumbs. Cover. Freeze 6 hours or overnight.

To Serve: Let stand at room temp. 10 - 15 minutes before serving.

Serves 20

CREAMY STRAWBERRY BLINTZES

Crêpes (recipe follows)
4 oz. cream cheese, softened
1½ tsp. vanilla
1 cup part-skim ricotta cheese
1½ cups chopped fresh
  strawberries

½ cup plus 2 tbsp. of no-
  sugar-added strawberry
  fruit spread
Sour cream or crème fraîche
Fresh mint leaves
Strawberry halves

Prepare crêpes; set aside. Preheat oven to 350°F. Place cream cheese and vanilla in food processor or blend container; cover and process until smooth. Add ricotta cheese; process until smooth. Stir in chopped strawberries. Spoon 2 heaping tbsp. filling down center of each crêpe and roll up. Place seam side down in lightly oiled or buttered 12 x 8-inch baking dish. Bake 15 to 18 minutes or until thoroughly heated. Serve warm with fruit spread. Top with strawberry halves and garnish with sour cream and mint leaves.

Makes 5 servings.

CRÊPES (FOR BLINTZES)

¾ cup all-purpose flour
¾ cup milk

1 tbsp. butter or margarine
2 eggs

continued next page

115

(Crêpes for Blintzes) ...Combine all ingredients in food processor or blender container; cover and process until smooth. Let stand at room temperature 1 hour or cover and refrigerate up to 8 hours. Process to combine again just before cooking. Heat 5 or 6" crêpe pan over medium heat. Lightly brush with oil or melted butter. Pour ¼ cup batter into hot pan all at once, tilting and rotating pan to spread batter evenly. Cook until bottom of crêpe is lightly browned. Turn over and continue cooking 30 seconds. Remove each crêpe to separate sheet of waxed paper. Repeat with remaining batter.

Makes 10 crepes.

RASPBERRY SUPREME DESSERT

Louise Hartung/Dorothy Skerritt, Langley.

2½ cups graham cracker crumbs	¾ cup icing sugar
½ cup butter, melted	8 oz. Cool Whip, thawed
2 tbsp. sugar	6 oz. Jell-O crystals, rasp.
8 oz. cream cheese, softened	1½ cups boiling water
	2 - 10 oz. raspberries, frozen

Combine butter, crumbs and sugar. Press into 9-inch x13" cake pan. Set aside. With electric mixer, blend cream cheese and icing sugar until creamy. Fold in thawed Cool Whip. Spread mixture over cracker crust. In large bowl dissolve gelatin in boiling water, add frozen raspberries. Stir gently and allow to set a few minutes until slightly thickened. Spread over cream cheese layer.

Refrigerate until ready to serve.

"It's pointless to go on asking, 'Why me?' The only worthwhile question is, 'Where do I go from here?'"

FROZEN BERRY MOUSSE

Marge Bekker, Grand Forks.

Shell:

1½ cup flour   ½ cup brown sugar
¾ cup chopped pecans   ¾ cup melted butter

Mix together well and spread on a cookie sheet like a giant cookie. Bake at 350°F for 20 minutes. Cool and break up into small pieces. Reserve ⅓ of mixture for topping. Spread remainder over bottom of 9-inch x 13" pan or 10-inch round cheesecake pan.

Filling:

3 egg whites   2 tbsp. lemon juice
¾ cup sugar   2 boxes frozen berries
1 package Cool Whip, thawed

Combine all ingredients and beat with electric mixer on high speed until mixture is thick. Fold in Cool Whip. Pour over layer in pan, sprinkle on reserved cookie crunchies. Freeze. Cut in squares or wedges to serve.

FROZEN STRAWBERRY CREAM

2 (250g) cream cheese, softened ½ cup sour cream
¾ cup sugar   3 cups fresh strawberries,
1 container Cool Whip, light   crushed or 2 (300g) frozen
  or regular, thawed   unsweetened strawberries,
Blueberry sauce   thawed and drained

Line a 9x5-inch loaf pan with double thickness of wax paper. In a large bowl of electric mixer, beat cream cheese, sour cream and sugar until mixed; fold in berries.

continued next page

(Frozen Strawberry Cream) ...Gently fold topping into berry mixture; spread into pan. Cover and freeze overnight. One hour before serving, place in refrigerator. To serve, spoon sauce on dessert plates, top with sliced dessert.

Serves 10.

FROZEN BERRY DESSERT

Base and Topping:

1 cup flour	¼ cup brown sugar, packed
½ cup pecans or walnuts, chopped	½ cup butter, melted

Filling:

2 egg whites	1 cup sugar
2 tbsp. lemon juice	2 cups fresh berries, sliced
1 cup whipping cream, whipped	

Base and Topping: In a medium bowl, combine all ingredients and then spread onto a baking sheet. Bake at 350°F for 20 minutes, stirring occasionally. Press ²/₃ of crumbs into a 9 x 13-inch pan. Reserve remainder of crumbs for topping.

Filling: In a large bowl, combine egg whites, sugar and lemon juice. Beat until stiff peaks form. Beat in berries briefly. (berries should be slightly broken up.) Fold in whipped cream and spoon filling over base. Top with reserved crumbs. Freeze for at least 6 hours before serving.

Cut into squares.

"Longing for the past, dreaming about the future, we pass up the present."

RASPBERRY BAKED APPLES

B.C. Raspberry Growers Assoc.

3 cups raspberries, fresh or   ½ cup sugar
  frozen without sugar   1 tbsp. minute tapioca
⅓ cup water   6 tart apples, medium
Whipped cream, plain or lightly sweetened

Mix together raspberries, sugar, tapioca, and water in large bowl. Quarter and core apples. Pare a lengthwise or vertical strip of peel from the center of each apple quarter. Rinse apples and (as they are pared) add to raspberry mixture. Stir to coat apples with raspberry mixture. Place in 7"x 11"baking pan. Cover with foil and bake for 350°F for about one hour or until apples are tender. While baking, stir apples occasionally (about once every 15 minutes) and spoon raspberry marinade over apples. Serve warm with cream or chilled with whipped cream. Serves 8.

BLUEBERRY COINTREAU

Meringues:

4 egg whites, at room   1½ cups sugar
  temperature   1 tsp. lemon juice

Filling:

1 pkg. (3¾ oz.) vanilla   1 cup milk
  pudding and pie filling   1 cup heavy cream
2 cups fresh or dry-pack   2 tbsp. cointreau
  frozen blueberries,   Orange peel curls
  rinsed and drained

Meringues: In a bowl, beat egg whites with an electric mixer until stiff. Gradually beat in sugar, ¼ cup at a time. Beat in lemon juice until stiff and glossy. Line a cookie sheet with foil. Spoon meringue mixture into 12 mounds on top of foil. Depress each mound, forming a cup-like shape. Bake at 275°F for 40 minutes. Cool in oven.

continued next page

(Blueberry Cointreau) ...**Filling:** Combine in a saucepan the pudding mix, milk, cream and liqueur. Stir over low heat until pudding bubbles and thickens. Cool to room temperature and fold in blueberries. Cover and chill. When ready to serve, place meringues on serving platter and fill each with blueberry filling. Garnish with orange peel curls and serve at once.

Serves 12.

RASPBERRY BOMBE

The name of this elegant dessert is French and means literally 'bomb'. Modern 'bombes' are classically molded in tall, conical molds or in melon-shaped molds. 'Bombes' are festive desserts and should be brought to the table in their entirety to be admired by guests before serving.

6 cups raspberry sherbet	2 tsp. vanilla extract
1 cup vanilla ice cream	1 tbsp. cointreau or
¾ icing sugar	grand marnier liqueur
Fresh raspberries	

Chill a 2 quart mold and line quickly with sherbet. Freeze. Force enough berries through a sieve to make 1 cup. Mix ice cream, sugar, vanilla, liqueur and berry juice. Fill center of mold with this cream mixture and freeze until firm. Unmold on serving plate, garnish with whole berries and whipped cream if desired.

Makes 6 - 8 servings.

SPICED JELLIED BLACKBERRIES

3 oz. berry-flavored gelatin	½ cup fresh orange juice
1½ cups boiling water	1¼ cup fresh blackberries
Dash *each* of ground cinnamon, nutmeg and cloves	

continued next page

(Spiced Jellied Blackberries) …Dissolve gelatin in boiling water. Add spices and orange juice. Chill until of the consistency of unbeaten egg whites. Stir in berries and chill until set. Serve with custard sauce or whipped or plain cream.

Makes 4 - 6 servings.

BLACKBERRY ROLY-POLY

Filling:
¼ cup all-purpose flour
Pinch of salt
Juice of 1 lemon
4 cups fresh blackberries
Sugar
2 tbsp. butter, melted
½ tsp. cinnamon

Biscuit Dough:
3 cups all-purpose flour
3 tsp. baking powder
1½ tsp. salt
½ cup shortening
1 cup milk (about)

Whipped cream

Biscuit dough: Sift dry ingredients. Cut in shortening until mixture resembles coarse cornmeal. Add enough milk to make a soft dough.

Filling: Mix flour, salt, lemon juice, 3½ cups blackberries and ¾ cup sugar. Roll biscuit dough on floured board to form a rectangle 16" x 9-inch. Brush with butter and sprinkle with 2 tbsp. sugar and the cinnamon mixed together. Place blackberry filling on center of dough, roll up and shape in a ring on a greased cookie sheet. Pinch ends. Make cuts in ring at 2" intervals, not cutting completely to center. Bake at 425°F for 25 - 30 minutes or until browned. Heap reserved berries in center and top with whipped cream and a berry. Serve at once.

Makes 6 - 8 servings.

"If you want your children to keep their feet on the ground, put some responsibility on their shoulders."

Preserving

NATURAL SUMMER FRUIT JAM

Bernardin of Canada, Ltd.

Recipe	Straw	Rasp	Blue	Red Cur
Tart Apples (Granny Smith)	5 med.	5 med.	5 med.	5 med.
Water	2 cups	2 cups	2 cups	2 cups
Fruit	8 cups	4 cups	4 cups	6 cups
Other	1 lemon	1 lemon	2 limes	1 lemon
Sugar	5½ cups	5 cups	3 cups	5½ cups
Boil Time (Approx.)	20 mn.	15-20 mn.	15-20 mn.	15-20 mn.
Yield	8½ pint	6½ pint	5½ pint	7½ pint

Directions: Natural Summer Fruit Jam

Wash and chop apples (including cores). Place in large, deep saucepan with water. Remove thin, yellow peel from lemon in one long piece. Squeeze juice. Set both aside. Chop remaining lemon pulp and add to apples. Bring to boil. Cover and simmer 20 minutes.

Prepare jars and canner. Pour cooked apples and liquid into a fine sieve. Using back of metal spoon, force as much applesauce through sieve to yield 2 cups. Return to saucepan.

Wash, prepare fruit. Add to applesauce with peel. Bring to a boil. Maintain constant boil while stirring in sugar until completely dissolved and avoid scorching. Boil vigorously until mixture reaches gel stage.

Prepare lids. Add reserved lemon (or lime) juice. Boil 1 minute. Remove from heat and skim off foam. Discard peel if desired.

Bottle, seal and process 5 minutes. Wipe jars, label and store.

APRICOT RED CURRANT JAM

Bernardin of Canada, Ltd.

"Prepare the individual fruits in season, then freeze to prepare this recipe later. Defrost thoroughly first."

2.2 lbs. apricots	4 cups stemmed red currants
1 large lemon	7 cups sugar

Select fruit which is ¾ ripe and ¼ slightly underripe. Wash, pit and chop apricots to make 5 cups (1.25 L). Remove stems and wash currants*. Cut zest (thin yellow peel) in one long strip from lemon and squeeze juice. Combine apricots, currants, lemon zest and juice in a large deep stainless steel or enamel saucepan. Bring to a boil. Maintaining boil, gradually stir in sugar. Stirring occasionally to prevent scorching, boil vigorously about 15 minutes or until mixture reaches gel stage. Remove from heat and skim off foam. Bottle, seal, process and store.

*Red currants have seeds which some people find objectionable in jams. To make a seedless jam, stem and wash currants shaking off excess moisture. Bring to a boil, cook and mash until tender, about 5 minutes. Pour mixture through a fine or cloth lined sieve. Use back of spoon and squeeze cloth to force as much mixture as possible through sieve. Combine sieved currants with apricots, lemon zest and juice and proceed as above. Yields: 7 - ½ pint jars.

Brandied Version: Pour 1 tbsp. Peach Apricot Brandy into each jar prior to adding hot jam. Fill and process as above.

"Discipline doesn't break a child's spirit half as often as the lack of it breaks a parent's heart."

BLACKBERRY JAM

Peel and seeds of 1 lemon   3 lbs. blackberries, hulled
3 tbsp. water   2 tbsp. lemon juice
3 lbs. granulated sugar

Tie lemon peel and seeds in a cheesecloth bag. Put it into a sauce-pan along with blackberries, water and lemon juice. Bring to a boil and simmer for 45 minutes or until the berries are cooked.

Add the sugar, stirring frequently, and as soon as it dissolves take the cheesecloth bag out and squeeze the pectin out of it into the jam. Turn up the heat and cook rapidly until the setting point is reached. Remove from heat and allow the jam to stand for 10 minutes, then spoon it into warm sterilized jars. Bottle, seal, process and store.

GREEN GOOSEBERRY JAM

4 lbs. gooseberries   3 cups water
4 lbs. granulated sugar

Cut stalks and tails off the gooseberries. Put the fruit into a sauce-pan with the water and slowly bring to a boil. Simmer until the gooseberries are very tender. Then add the sugar and stir until it dissolves. Bring to a rolling boil for about 15 to 20 minutes or until the setting point is reached. Remove from the heat and let the jam stand for 15 minutes. Bottle, seal, process and store.

"Unless we replace yesterday's feelings with today's experiences, we will remain bound by the past when we could be free in the present."

RED CURRANT & ORANGE JAM

2 lbs. red currants, stalked
2 lbs. granulated sugar

2 oranges, finely sliced

Put the red currants and orange slices into a saucepan. Bring slowly to a boil and cook gently for 10 minutes. Add the warmed sugar and bring slowly to a boil again. Boil rapidly for 7 to 10 minutes or until the setting point is reached. Remove from the heat and let the jam stand for 15 minutes. Spoon into sterilized jars. Bottle, seal, process and store.

RASPBERRY JAM

2 lbs. raspberries, crushed   1½ lbs. granulated sugar

Combine raspberries and sugar in a saucepan and bring slowly to a boil, stirring constantly. The raspberry juice will soon begin to flow. When there is enough juice, boil rapidly until the jam is thick and the setting point is reached. Allow to stand off the heat for 10 minutes, stirring the jam once. Spoon it into hot sterilized jars. Bottle, seal, process and store.

BEST FROZEN RASPBERRY JAM

Marge Bekker, Grand Forks

"I have made this freezer jam for 30 years and we all love it! It takes less sugar than most recipes."

¼ cup sugar
2 cups mashed raspberries
4 tsp. lemon juice

1 pkg. fruit pectin crystals
2 cups sugar

continued next page

(Best Frozen Raspberry Jam) ...3½ to 4 cups raspberries will make 2 cups crushed. In large mixer bowl, mix ¼ cup sugar and pectin crystals. Add crushed fruit and beat 7 minutes at low speed. Add 2 cups sugar and lemon juice. Beat 3 minutes longer. Pour into freezer containers, cover and let stand 24 hours to thicken. Freeze. When thawed, store in refrigerator. Makes about 2 pints.

STRAWBERRY & RED CURRANT JAM

3 lbs. strawberries, hulled   1½ lbs. red currants, cleaned
3 lbs. granulated sugar

Put the strawberries and red currants into a saucepan and let them cook over a low heat. Meanwhile, warm the sugar in the oven. Keep stirring fruit as the juice starts running. When fruits are cooking in their own juice, add the warmed sugar and stir until sugar is dissolved. Bring the jam to a rolling boil, stirring frequently until it thickens and reaches the setting point. It should take 15 to 20 minutes.

Leave jam for 10 minutes off the heat before spooning it into hot sterilized jars. Seal, process and store.

TROPICAL JAM

6 lbs. fresh strawberries   4 oz. can crushed pineapple,
2 tbsp. lemon juice   drained
7¼ cups sugar   6 oz. liquid pectin

Mash strawberries and measure 3½ cups. In large saucepan, combine berries, pineapple, lemon juice and sugar. Mix well. Over high heat, bring to a full, rolling boil. Boil 1 minute, stirring constantly. Remove from heat. Add pectin. Stir and skim off foam for 15 minutes. Pour into hot, sterilized jars. Seal, process and store.

MOM'S GOOSEBERRY JAM

4 cups prepared gooseberries   6 cups sugar
½ bottle fruit pectin

Chop gooseberries in blender. Measure 4 cups into very large sauce-pan. Add sugar and mix well. Put over high heat. Bring to a full rolling boil and boil hard for 1 minute, stirring constantly. Remove from heat and stir in pectin at once. Skim off foam with metal spoon. Skim and stir for 5 minutes to cool and help prevent floating fruit. Ladle into hot sterilized jars.

Seal, process and store.

MICROWAVE STRAWBERRY JAM

1 lb. strawberries, hulled   1 tbsp. lemon juice
1½ cups granulated sugar

Place strawberries and lemon juice in a large bowl, at least 12-cup capacity. Cover and cook on high for 5 to 6 minutes, until soft. Add the sugar and stir gently until dissolved. Cook on high, uncovered, for 12 to 15 minutes, until the setting point is reached. Test for setting point frequently after 10 minutes. Cool slightly. Stir and pour into hot sterilized jars.

Seal, process and store.

"Often the best way to lead others is by assuring them that we are right behind them."

SUGARLESS RASPBERRY JAM

B.C. Raspberry Growers Assoc.

1 tbsp. gelatin
1½ to 2 tbsp. liquid artificial
sweetener

2 cups crushed raspberries
1 tbsp. lemon juice

Soften gelatin in ½ cup of crushed raspberries. After 5 minutes, dissolve over hot water pan. Combine gelatin mixture, raspberries, lemon juice and desired amount of liquid sweetener and mix well. Pour into hot sterilized jars and seal.

This jam must be refrigerated and will keep up to 6 weeks.

SPICED BLUEBERRY JAM

6 cups fresh blueberries
2 tbsp. lemon juice
¼ tsp. cloves
¼ tsp. cinnamon

¼ tsp. allspice
1¾ oz. powdered pectin
5 cups sugar

Purée blueberries. Measure 4 cups of purée. In large saucepan, combine purée, lemon juice, spices and sugar. Stir to mix well. Over high heat, bring to a full, rolling boil. Boil 1 minute, stirring constantly. Remove from heat. Stir in pectin. Stir and skim off foam for 5 minutes. Pour into hot, sterilized jars. Seal, process and store.

Yields: 8 - 6 oz. jars.

"The questions that keep us searching may be more captivating than the answers we seek."

BLUBARB AND RASPBERRY JELLY

Bernardin of Canada, Ltd.

3 cups washed blueberries
1 cup rhubarb
2 cups sugar
1 pouch (85 ml.) liquid pectin

2 cups raspberries
2 cups water
1½ cups honey

In medium stainless steel or enamel saucepan, combine berries, rhubarb and water. Cook gently 5 minutes, mashing berries. Remove from heat. Pour through a dampened cheesecloth lined strainer or jelly bag. Allow to drip 8 hours or overnight. Fill boiling water canner with water. In stainless steel or enamel saucepan, combine 2 cups juice, sugar and honey, stirring to dissolve sugar. Bring to a full rolling boil over high heat, stirring constantly. Add liquid pectin; return to a boil; stir and boil hard 1 minute. Remove from heat; skim off foam with metal spoon. Bottle, seal, process and store.

Yields: 5 - ½ pint jars.

BLACKBERRY & APPLE JELLY

2 lbs. blackberries
4 tbsp. lemon juice
Granulated sugar

1 lb. cooking apples
2 cups water

Hull the blackberries. Peel, core and chop the apples. Put the fruit, lemon juice and water into a saucepan and simmer until tender. Mash the fruit and strain it through a jelly bag or layers of cheesecloth. It will take 1 to 2 hours to strain. Do not touch the jelly bag. Measure the juice by the cupful, and set aside the same amount of sugar. Bring the juice to a boil and continue until it thickens a little, then add the sugar. Stir well and keep boiling rapidly until the setting point is reached. Immediately ladle into hot sterilized jars. Seal, process and store.

FRESH CURRANT JELLY

4 lbs. fresh ripe currants   7 cups sugar
1 cup water   ½ bottle fruit pectin

Crush currants. Add water and bring to a boil. Simmer covered for 10 minutes. Put in jelly cloth bag and let hang to extract juice. Measure 5 cups of juice into a very large saucepan. Add sugar and mix well. Put over high heat and bring to boil, stirring constantly. Stir in pectin at once. Bring to a full rolling boil and boil hard for 1 minute, stirring constantly. Remove from heat, skim off foam. Bottle, seal, process and store.

OLD FASHIONED RED CURRANT JELLY

4 lbs. red currants   2 cups water
Sugar

Wash the red currants and place in a preserving pan with the water. Bring to a boil. Reduce the heat and simmer for 45 minutes to an hour until the fruit is very soft.

Mash well and strain through a jelly bag. Measure the juice. Return the juice to a clean pan and bring to boil. Reduce the heat and add 1 lb. sugar to each 2 cups juice. When the sugar has dissolved, boil rapidly for 10 minutes until setting point is reached. Bottle, seal, process and store. Note: for a thicker jelly to be used as a condiment, the fruit may be cooked with no water. After straining add 1¼ lb. sugar to 2 cups hot juice. The yield will be slightly less.

"We must learn to enjoy the sweet moments in our lives, to endure the bitter ones, and to make the most of the disruptive transitions."

GOOSEBERRY JELLY

4 lb. gooseberries   6 cups water
Sugar

Wash the gooseberries and place in a preserving pan with the water. Bring to a boil. Reduce the heat and simmer for 45 minutes to an hour until the fruit is very soft. Wash well and strain through a jelly bag. Measure the juice and add 1 lb. sugar to each 2 cups of juice. When the sugar has dissolved, boil rapidly for about 10 minutes until setting point is reached. Ladle into hot sterilized jars. Seal, process and store.

RASPBERRY & RED CURRANT JELLY

4 cups raspberries   8 cups stemmed red currants
Sugar

In a large heavy saucepan, crush raspberries thoroughly. Add currants in batches. Crush each addition to start juice flowing. Bring to a slow boil over medium heat. Simmer gently, covered, until fruit is soft and seeds look clean (about 10 minutes). Mash once more and simmer for 2 minutes longer. Transfer to dampened jelly bag. Let drip for 24 hours. Measure juice into a large heavy saucepan. Stir in an equal quantity of sugar and mix well. Bring to a full rolling boil over high heat. Boil vigorously until jelly sets (about 8 minutes) stirring frequently. Bottle, seal, process and store.

"Beware of those times when your way of seeing things seems like the only truth."

APPLE STRAWBERRY JELLY

1½ cups unsweetened   3½ cups sugar
  bottled apple juice   1 pouch liquid pectin
2 cups fresh strawberries, mashed one layer at a time

Pour apple juice in a large saucepan. Measure ½ cup prepared berries, add to apple juice. Add sugar. Bring to a boil over high heat, stirring constantly. At once stir in pectin. Bring to a full rolling boil. Boil hard for 1 minute, stirring constantly. Remove from heat. Skim off foam. Pour into hot sterilized jars. Seal, process and store.

Yields: 3½ cups.

BLUEBERRY MARMALADE

1 large orange, quartered   1 medium lemon, quartered
2 cups water   2 cups fresh blueberries,
1 box fruit pectin crystals   washed and dried
6 cups sugar

Slice orange and lemon quarters thinly into a preserving kettle. Include juices and add water. Cover kettle. Bring mixture to a boil over high heat. Reduce heat slightly and simmer for 30 minutes.

Crush blueberries and add to kettle. Add pectin crystals and bring to a hard boil over high heat, stirring continuously. Stir in sugar all at once. Bring mixture to a full rolling boil. Boil hard for 1 minute, stirring continuously. Remove kettle from heat. Cool for about 7 minutes, skimming off foam with a metal spoon and stirring occasionally. Ladle into hot sterilized jars. Seal, process and store.

Yields: 9 - ½ pint jars.

"It's not easy to admit we're capable of doing what we want to believe we aren't."

SPIRITED FRUIT

Bernardin of Canada, Ltd.

For Blackberries:

12 cups blackberries
1 ½ tsp. creme de cassis
  or marsala per jar

1 tbsp. rum, brandy or vodka
  per jar

For Blueberries:

12 cups blueberries
1 ½ tsp. Grand Marnier per jar

1 tbsp. rum, brandy, vodka
  per jar

For Syrup:

2 cups water

1 cup sugar

Wash berries in cold or ice water to firm fruit, then drain. Rinse and drain blueberries. Prepare syrup in a large stainless steel or enamel saucepan. Combine sugar and water as indicated above. Bring to a boil. Add fruit to hot syrup. Return mixture to a boil and boil gently for 5 minutes. Pack hot fruit in hot jar to within ¾ inch of top rim. Pour liquor or liqueur over fruit in jar. Add hot syrup to cover fruit, leaving ½ inch head space. Bottle, seal, process and store. Makes 7 half-pint jars.

SPREADABLE STRAWBERRIES

Bernardin of Canada, Ltd.

12 cups strawberries
5 tart green apples

2-355ml undiluted frozen
  apple juice, thawed

Place strawberries in a large, stainless steel or enamel saucepan. Peel, core and finely chop apples. Add to strawberries with apple juice concentrate. Stirring occasionally, bring mixture to a boil. Reduce heat; stirring frequently to prevent sticking. Boil gently about 30 minutes or until thickened to spreadable consistency. Bottle, seal, process and store. Makes 7 half-pint jars.

VERY BERRY SPREADABLE FRUIT

Bernardin of Canada, Ltd.

"For optimum flavor use only locally grown, fresh-picked fruit. (Imported fruits generally do not have sufficient flavor intensity.) If unable to prepare and process during the peak season, prepare and measure fruit for recipe and freeze. Thaw fruit and prepare recipe at a later date."

6 cups washed and hulled strawberries
4 tart green apples

3 cups pitted cherries
3 cups raspberries
2-355ml undiluted frozen apple juice, thawed

Use ¾ of just ripe fruit and ¼ slightly underripe fruit. Place strawberries in a large, stainless steel or enamel saucepan. Peel, core and finely chop apples. Add to strawberries with cherries, raspberries and apple juice concentrate. Stirring occasionally, bring mixture to a boil. Reduce heat, stirring frequently to prevent sticking. Boil gently about 45 minutes or until thickened to spreadable consistency. Bottle, seal, process and store. Makes 7 half-pint jars.

DIABETIC STRAWBERRY SPREAD

Louise M. Wunderlich, Langley

2 cups sliced fresh or frozen unsweetened strawberries
Artificial sweetener equivalent to 8 to 12 tsp. of sugar

¼ cup water
1½ tsp. unflavored gelatin
Red food coloring (optional)

Sprinkle gelatin over water to soften. Let stand 5 minutes. Place berries in a medium saucepan; bring to a boil. Cook, stirring occasionally for 5 minutes. Add softened gelatin, sweetener and food coloring. Stir until gelatin dissolves. Skim foam from surface. Bottle, seal, process and store.

STRAWBERRY RHUBARB CONSERVE

Bernardin of Canada, Ltd.

6 cups strawberries
2 cups finely chopped rhubarb
¼ cup lemon juice

1 orange, washed
1 cup raisins
1 cup coarsely chopped
  pecans, optional

Rinse strawberries in cold running water. Remove stems, drain and pat dry. Cut strawberries in half. Quarter unpeeled orange and chop finely by hand or in food processor. Recording the number of cups, measure strawberries, orange, rhubarb, raisins and lemon juice into a large heavy stainless steel or enamel saucepan. Place mixture over medium-high heat. Gradually stir in ¾ cup sugar for each cup fruit. To prevent scorching, continue stirring until mixture boils. Boil uncovered. Stir occasionally, until mixture reaches gel stage, 30 to 40 minutes. If using nuts, stir them into thickened fruit mixture. Boil 1 minute longer. Remove from heat. Bottle, seal, process and store. Yields: 8 - ½ pint jars.

STRAWBERRY VINEGAR

6 cups fresh strawberries
6 tbsp. sugar, or to taste

3 cups white wine vinegar or
oriental rice vinegar

Sort berries, rinse, drain well, and remove hulls. Crush or chop berries and combine with vinegar in a sterilized, dry two-quart jar or crock. Cover the container closely and let mixture stand for a month, shaking or stirring it occasionally.

Empty the mixture into a fine-meshed sieve lined with fine nylon net or 2 layers of dampened cheesecloth and set over a bowl. Let the vinegar drain, pressing lightly with the back of a spoon to obtain the last of the juice. Discard remaining pulp.

continued next page

(Strwaberry Vinegar) ...Combine the vinegar and sugar in a stainless-steel or enameled saucepan and heat it just to simmering. Simmer uncovered for 3 minutes. Remove from heat and let cool completely. Skim off foam and strain vinegar into hot sterilized jars. Seal, process and store.

RASPBERRY WINE VINEGAR

Lorraine Bolton, Vancouver.
In loving memory of my father, John Bolton. He did not take
alcoholic beverages, but was very fond of Raspberry Vinegar.

5 lb. fresh raspberries   4 cups cider/wine vinegar
1 cup sugar to each 2 cups of liquid

Crush berries. Cover berries with vinegar and let stand for 4 days, stirring occasionally. Strain through 3 layers of cheesecloth. Add sugar and boil for 15 minutes. Bottle into hot sterilized jars. Seal, process and store.

To Serve: As a drink: Mix 1 tbsp. of raspberry vinegar with hot or cold water.

RASPBERRY VINEGAR

Jean Gregory, Burnaby.

4 lbs. fresh whole raspberries   8 cups white sugar
6 cups white vinegar

Wash raspberries, place in crock. Pour vinegar over to cover and let stand 1 week. Hang in three layers of cheesecloth for two days to strain. Add sugar and boil for 1½ hours. Bottle into hot sterilized jars. Seal, process and store.

To Serve: Mix with hot water or soda as a drink. Or can be used in various recipes.

136

Miscellaneous

BLUEBERRY-ORANGE BREAD

2 tbsp. butter
¼ cup boiling water
⅔ cup orange juice
4 tsp. orange rind, grated
1 egg
1 cup sugar

2 cups sifted all-purpose flour
1 tsp. baking powder
¼ tsp. baking soda
½ tsp. salt
1 cup blueberries, fresh or
    frozen, thawed
2 tbsp. honey

Melt butter in boiling water in small bowl. Add ½ cup orange juice and 3 tsp. rind. Beat egg with sugar until light and fluffy. Add sifted dry ingredients alternately with orange liquid, beating until smooth. Fold in berries. Bake in greased fancy 1½ quart baking dish or loaf pan in 325°F oven for about 1 hour and 10 minutes. Turn out on rack or tray. Mix 2 tbsp. orange juice, 1 teaspoon rind, and honey. Spoon over hot loaf. Let stand until cold.

Makes 1 loaf.

MRS. FLAHERTY'S BLUEBERRY LOAVES

6 eggs
4 cups sugar
1½ cups milk
1½ tbsp. ground cardamom

8 cups all-purpose flour
10 tsp. baking powder
4 cups blueberries, fresh or
    frozen, thawed and drained

In a large bowl, mix eggs and sugar. Stir in milk and cardamom. Stir in flour and baking powder. Fold in blueberries. (If desired, blueberries may be mixed with a little flour before adding to dough.) Spoon mixture into 4 greased and floured loaf pans. Bake in pre-

continued next page

Flaherty's Blueberry Loaves) heated moderate oven (350°F) for 55 minutes to 1 hour or until center feels firm to the touch.

…Unmold and cool on rack. Wrap and store in freezer until needed. Loaves may be decorated with frosted topping, sugar topping or with coconut topping.

Makes 4 loaves.

RASPBERRY PICNIC BREAD

Sheila Ross, N. Vancouver

"My father grows raspberries with great success in his backyard in North Vancouver. He freezes the excess and then makes jam throughout the winter (the best raspberry jam in the world!). This recipe is a good way to use up the extras."

1¼ cups raspberries, frozen thawed and undrained	1½ cups flour
1 tbsp. raspberry jam	1 cup sugar
2 eggs	1 tsp. cinnamon
¾ cup vegetable oil	½ tsp. baking soda

In food processor or blender, purée raspberries and jam. In a small bowl, combine eggs and oil using wire whisk. Sift flour, sugar, cinnamon, and baking soda into second bowl. Make a well in center of dry ingredients. Pour in egg mixture and raspberry purée. Blend well. Pour batter into greased 9x5" loaf pan. Bake in preheated 350°F oven for 50 to 60 minutes, or until tester inserted in center comes out clean. Cool in pan 5 minutes. Remove to rack Cool completely before slicing.

"Emotional bonding comes through the hundreds of 'little' things we do for our children."

STRAWBERRY BREAD

1 cup butter
1½ cups sugar
4 eggs
3 cups flour
¾ tsp. cream of tartar
½ tsp. baking soda

1 cup strawberry jam
½ cup sour cream
1 tsp. vanilla
1 tsp. lemon juice
1 tsp. grated orange peel
½ cup chopped nuts

In a large bowl, cream butter and sugar. Beat in eggs, one at a time, until mixture is light and fluffy. Combine dry ingredients and set aside. Combine remaining ingredients except nuts and add alternately with dry ingredients to butter mixture. Stir in nuts if desired. Pour into 2 greased loaf pans. Bake at 350°F for 60 minutes or until done. Cool before removing from pan.

Makes 2 loaves.

HOMEMADE BLUEBERRY WAFFLES

2 eggs whites
2 egg yolks
1½ cups milk
½ cup butter, melted
1 cup fresh blueberries

2 cups sifted all-purpose flour
2 tsp. baking powder
¾ tsp. salt
1 tbsp. sugar

Beat egg whites and set aside. Beat egg yolks and add milk and butter. Sift dry ingredients and add to egg mixture. Mix until smooth. Fold in egg whites and berries. Bake in hot waffle iron. Serve with syrup or hot blueberry sauce.

Makes 6 waffles.

"Your child will never grow too old to hear you say, 'I love you'."

HOMEMADE BLUEBERRY PANCAKES

1½ cups all-purpose flour
2½ tsp. baking powder
3 tbsp. sugar
¾ tsp. salt
2 egg yolks

1 cup milk
3 tbsp. butter, melted
1 cup fresh blueberries
2 egg whites, stiffly beaten

Sift dry ingredients. Beat egg yolks and combine with milk and butter. Add to dry ingredients and mix until smooth. Stir in berries. Fold in stiffly beaten egg whites. Bake on hot greased griddle. Serve with blueberry sauce.

Makes 6 pancakes.

STRAWBERRIES AND CHEESE

1 cup creamed cottage cheese
¼ cup cereal cream
¼ cup plain yogurt
2 tsp. orange rind, grated

½ tsp. fresh ginger, grated
2 - 3 cups fresh strawberries
Mint sprigs for garnish

Blend cottage cheese, cereal cream and yogurt until smooth. Scoop into bowl and stir in orange rind and ginger. Line a small strainer with cheesecloth (triple thickness) and suspend over bowl. Spoon in the cheese mixture, cover and refrigerate at least 12 hours. Turn onto serving platter and remove cheese cloth. Arrange strawberries, halved and garnish with mint sprigs.

"A child enters your home and makes so much noise you can hardly stand it. Then the child departs, leaving the house so quiet you think you'll go mad."

BERRY SHAKE

½ cup shelled almonds (soaked 5 hours in one cup water and drained)
½ cup pitted dates, (soaked 1 hour in 2½ cups water)

$^1/_3$ cup frozen strawberries
$^1/_3$ cup frozen blueberries
$^1/_3$ cup frozen red or black raspberries

Blend ingredients together at high speed for one minute. Strain through cheesecloth or strainer, collecting liquid in large bowl. Pour liquid into pitcher. Serve immediately for best color, taste and nutritional value. Otherwise, store in covered jar in refrigerator. The sooner used, the better the flavor.

Stir or shake before pouring into glasses.

BLUEBERRY MILKSHAKE

6 oz. vanilla ice cream
3 oz. milk
2 oz. blueberry sauce

Whipping cream (optional)
Fresh blueberries

Combine ice cream, milk and blueberry sauce in cup of fountain blender. Blend on high until thick and creamy. Pour into serving glass. Garnish with a dollop of whipped cream and additional fresh blueberries.

Makes 1 serving.

FRESH CURRANT ICE

1 quart fresh currants
1 tsp. unflavored gelatin
Juice of ½ lemon

2¼ cups water
1 cup sugar
$^1/_8$ tsp. salt

continued next page

141

(Fresh Currant Ice) …Wash currants, remove stems, and put in saucepan with 2 cups water. Bring to boil and simmer for 5 minutes, or until currants are soft. Force through fine sieve and bring to boil. Sprinkle gelatin on remaining ¼ cup water to soften; add hot currant pulp and stir until dissolved. Stir in remaining ingredients. Pour into refrigerator tray and partially freeze. Transfer to chilled bowl. Beat with a rotary beater until fluffy and light in color. Return to tray and freeze until firm. Makes 6 servings.

ITALIAN STRAWBERRY ICE

2 quarts strawberries	1 cup sugar
1 cup water	Juice of 1 small lemon

Wash and hull berries; then purée in a blender. Boil sugar and water together for 5 minutes. Cool. Combine with berry purée and stir in lemon juice. Pour into refrigerator trays and freeze to a mush, stirring occasionally.

Makes 4 - 6 servings.

BLACK CURRANT LIQUEUR

For Making Kir

1 lb. fresh black currants	1 cup sugar
2½ cups brandy	3-inch cinnamon stick
1 clove	

Crush black currants with a fork and combine with remaining ingredients. Place in a large preserving jar. Store in a dark place for one month or so. Strain through a sieve lined with cheesecloth. Rebottle. Makes 2½ cups

(Blackcurrant Liqueur)...**Kir:** Add a splash of black currant liqueur to a glass of chilled dry white wine.

Kir Royale: Add a splash to a glass of chilled champagne.

Kir Cardinal: Add a splash to a glass of red wine.

STRAWBERRY PARTY PUNCH

Great for summertime festivities!

2 cups fresh strawberries
   w/stems
3 oz. pkg. gelatin, straw.
   flavored juice, thawed
1 cup boiling water

12 oz. can frozen pink
   lemonade, thawed
6 oz. can frozen orange
6 cups cold water
   3½ cups ginger ale,
      chilled

Wash, dry and freeze strawberries with stems on cookie sheet, one layer deep. Freeze until firm.

In small bowl, dissolve gelatin in boiling water. Cool. In large punch bowl, combine gelatin mixture, juice concentrates and cold water. Just before serving, gently stir in ginger ale and add the frozen strawberries. Note: A few frozen blueberries would add colour.

Makes 28 - ½ cups.

GOLDEN PUNCH

A Party Favorite

Ice Ring: Can be made several days in advance. Remove from the mold, wrap in plastic and store in freezer.
1-6 oz. can frozen lemonade concentrate, thawed
3 cups water
1 cup fresh strawberries with stems or frozen without sugar

continued next page

(Golden Punch) ...In a 2 quart non-metal container, combine lemonade and water. Mix well. Pour lemonade into a 6 cup mold. Freeze until slushy, about 2 - 3 hours. Place strawberries in slush, allowing tops of strawberries to show. Freeze until firm or overnight.

Golden Punch:

1 - 12 oz. can frozen grapefruit juice concentrate, thawed
1 - 12 oz. can frozen lemonade concentrate, thawed
7 cups cold water
4 cups ginger ale

In 6 quart non-metal container, combine grapefruit, lemonade concentrates and water. Mix well. Refrigerate until chilled. Just before serving, place ice ring, strawberry side up into punch bowl. Carefully pour punch over ring and add ginger ale. Stir gently. Makes 26 - ½ cup servings.

"When God thought of mother, He must have laughed with satisfaction—so rich, so deep, so divine, so full of soul, power and beauty was the conception!"

CHARTS AND CONVERSIONS

WEIGHTS AND MEASURES

1 tbsp. = 3 tsp.
¼ cup = 4 tbsp.
⅓ cup = 5 ⅓ tbsp.
½ cup = 8 tbsp.
⅔ cup = 10 ⅔ tbsp.
¾ cup = 12 tbsp.
1 cup = 16 tbsp.
1 oz. = 28.35 grams
.035 oz. = 1 gram

1 cup = 8 fl. oz.
1 cup = ½ pint
2 cups = 1 pint
4 cups = 1 quart
4 quarts = 1 gallon
8 quarts = 1 peck
4 pecks = 1 bushel
1 quart = 946.4 ml
.06 quarts = 1 litre

HOW MUCH AND HOW MANY

Butter, Chocolate:
  2 tbsp. butter = 1 oz.
  1 stick or ¼ lb. butter = ½ cup
  1 sq. chocolate = 1 oz.

Crumbs:
  28 saltine crackers = 1 cup fine
  14 sq. graham crackers = 1 cup fine
  22 vanilla wafers = 1 cup fine
  1½ slices bread = 1 cup soft
  1 slice bread = ¼ cup fine dry

Fruits:
  Juice of 1 lemon = about 3 tbsp.
  Juice of 1 orange = about ⅓ cup
  1 lb. whole berries = approx. ½ - ¾ cup prepared fruit
  1 med. apple, chopped = about 1 cup

continued next page

Cream, Cheese, Eggs:

1 cup whipping cream = 2 cups whipped
1 lb. cheese, shredded = 4 cups
8 egg whites = 1 cup
8 egg yokes = ¾ cup

INGREDIENT SUBSTITUTIONS

1 cup sifted all-purpose flour = 1 cup unsifted all-purpose flour
minus 2 tbsp.
= 1¼ cups sifted cake and pastry
flour

1 cup sifted self-rising flour = 1 cup sifted all-purpose flour
plus 1½ tsp. baking powder and
½ tsp. salt.

1 cup granulated sugar = 1 cup brown sugar, firmly packed

1 tbsp. cornstarch (for thickening) = 2 tbsp. flour
= 2 tsp. quick cooking tapioca

1 tsp. baking powder = ¼ tsp. baking soda plus ¾ tsp. cream of
tartar

1 cup butter = 1 cup margarine (hard or brick-type)
= 1 cup shortening

1 cup liquid honey = 1¼ cups sugar plus ¼ cup liquid

1 cup corn syrup = 1 cup sugar plus ¼ cup liquid

1 cup buttermilk or sourmilk = 1 tbsp. lemon juice or vinegar
plus enough milk to make 1 cup
(let stand 5 min.)

1 cup buttermilk or sour cream = 1 cup plain yogurt

1 cup milk = ½ cup evaporated milk plus ½ cup water

1 cup cream = ¾ cup milk plus ¼ cup butter

1 oz. chocolate (1 square) = 3 tbsp. cocoa plus 1 tbsp. butter

1 pkg. active dry yeast = 1 tbsp. active dry yeast

1 whole egg = 2 egg yolks

Juice of 1 lemon = 3 to 4 tbsp. bottled lemon juice

HOW MUCH TO BUY?

Note: The fruit used to measure in this table is NET weight. It had been sorted, washed and stemmed. Therefore, extra fruit must be bought to allow for sorting, removal of stems and the occasional berry that gets popped into your mouth!

Cups of Prepared Fruit
(Sorted, Cleaned and Crushed)

Berry	1 lb.	2.5 lb.	5 lb.	10 lb.	20 lb.
Strawberry	2	5	10	20	40
Blueberry					
Blackberry					
Raspberry	1.75	4.5	8.75	17.5	35
Gooseberry					
Black & Red Currant					

When determining how much fruit to buy, consider the following:

1) Have I purchased all the necessary ingredients to jam, jelly or cook beforehand?

2) Do I have all the necessary equipment (canner, juicer, jars, etc.) on hand?

3) Do I have enough time set aside to complete the task today, while the fruit is fresh?

4) Are my helpers lined up?

Enjoy the process, it can be a lot of fun! You will feel proud of your efforts as your family sing you praises as they enjoy the fruits of your labour in the wintertime.

SYRUP CHART

(For Canning and Freezing)

This syrup is great for canning all fruit and for making fresh fruit salad. There's nothing to it!

Instructions: Boil sugar and water until sugar dissolves. Chill.

Syrup	Sugar	Water	Yield
30% - Medium	2 cups	4 cups	5 cups
40% - Med/Heavy	3 cups	4 cups	5.5 cups
50% - Heavy	4.75 cups	4 cups	6.5 cups
60% - Extra Heavy	7 cups	4 cups	7.75 cups

Freezing:

Packed with syrup, requires 2" head space.

Packed without sugar or dry pack, requires 1" headspace.

Canning:

Raw Pack with syrup, fruit juice or water- 1½ inch headspace.

Hot Pack with syrup, fruit juice or water—½ inch headspace.

Processing:

Raw Pack—canned berries requires a Boiling Water Bath:
Pints—15 minutes Quarts—20 minutes
   Half Gallon Jars—30 minutes

There is nothing like the taste of fresh fruit all year 'round!

VOLUME MEASUREMENTS (DRY)

$^1/_8$ tsp. = 0.5 ml

$^1/_4$ tsp. = 1 ml

$^1/_2$ tsp. = 2 ml

$^3/_4$ tsp. = 4 ml

1 tsp. = 5 ml

1 tbsp. = 15 ml

2 tbsp. = 30 ml

$^1/_4$ cup = 60 ml

$^1/_3$ cup= 75 ml

$^1/_2$ cup= 125 ml

$^2/_3$ cup= 150 ml

$^3/_4$ cup= 175 ml

1 cup= 250 ml

2 cups (1 pint)= 500 ml

3 cups= 750 ml

4 cups (1 qt.) = 1 litre

VOLUME MEASUREMENTS (FLUID)

1 fl. oz. = 30 ml

4 fl. oz. = 125 ml

8 fl. oz. = 250 ml

12 fl. oz. = 375 ml

16 fl. oz. = 500 ml

WEIGHTS (MASS)

$^1/_2$ oz. = 15 grams

1 oz. = 30 grams

3 oz. = 90 grams

4 oz. = 120 grams

8 oz. = 225 grams

10 oz. = 285 grams

12 oz. = 360 grams

16 oz. = 450 grams

DIMENSIONS

$^1/_{16}$ inch = 2 mm

$^1/_8$ inch = 3 mm

$^1/_4$ inch = 6 mm

$^1/_2$ inch = 1.5 cm

$^3/_4$ inch = 2 cm

1 inch = 2.5 cm

"Taking risks by denying our fears is reckless. Taking risks by facing our fears is courageous."

MANUFACTURER'S JAM

The fruit used to calculate the number of cups of prepared fruit was washed, drained, sorted and stemmed. Therefore, you must add approximately ¼ to ½ lb. per recipe.

Manufacturer (1 batch=1pkg/bottle)	Fruit lbs.	Prepared (cups)	Sugar (cups)	Yield (cups)
STRAWBERRY				
Cooked:		Fruit		
Slim Set	3	6	4.5	8
Certo Regular	2.5	4.5	7	8
MCP	3	5.75	8.5	11
Certo Light	3	6	4.5	8
Certo Liquid	2	3.75	7	7.5
Freezer-No Cook:				
Certo Regular	1	2	4	5
MCP	2	3.25	4.5	7
Certo Liquid	1	1.75	4	5
Garden Fare	2	4	1.5	4.5
RASPBERRY				
Cooked:				
Slim Set	3.35	5.5	4	8
Certo Regular	3	5	7	8
MCP	3.5	6	8.5	10
Certo Light	3.5	6	4.5	8
Certo Liquid	2.25	3.75	6.5	8
Freezer-No Cook:				
Certo Regular	2	3	5.25	6.5
MCP	2	3.25	4.5	7
Certo Liquid	1.25	2	4	4.5
Garden Fare	2.5	4	1.5	4.5
Manufacturer (1 batch=1pkg/bottle)	Fruit lbs.	Prepared (cups)	Sugar (cups)	Yield (cups)

BLUEBERRY
Cooked:

Slim Set	3	6	3	8
Certo Regular	2	4	5	7
MCP	2	3.75	6	7
Certo Liquid (2 pouches)	2.25	4.5	7	9

Freezer-No Cook:

Garden Fare	1.5	4	1.5	4.5

BLACKBERRY
Cooked:

Slim Set	3	6	4.5	9
Certo Regular	2.5	5	7	8
Certo Liquid	2	3.75	7	7.5
MCP	3	5.75	8	10

Freezer-No Cook:

Garden Fare	2	4	1.5	4.5
Certo Liquid	1	2	4	4.5
MCP	1.75	3.25	4.5	7

GOOSEBERRY
Freezer:

Certo Liquid	2.25	3.75	6	8

CURRANT
Freezer:

Certo Liquid - Black	2.5	4	7.5	8
- Red	2.5	4	6.5	8

SUGARLESS JAM

Slim Set - Cooked:

Strawberry	2	4	sweetener 4
Raspberry-w/apple juice	2	3.5	sweetener 4
Blueberry-w/apple juice	2	4	sweetener 6
Blackberry-w/juice	2	4	6

MANUFACTURER'S JELLY

NOTE: More fruit is needed to make jelly than jam.

Manufacturer (1 batch=1pkg/bottle)	Fruit lbs.	Prepared (cups)	Sugar (cups)	Yield (cups)
STRAWBERRY				
Cooked:		Fruit Juice		
Slim Set	2.5	5	3.5	7
MCP	1.5	3	4.5	5
Certo Liquid (2 pouches)	2	4	7.5	8
RASPBERRY				
Slim Set	3	5	3	6
Certo Regular	2.5	4	5.5	7
MCP	2.5	4	5	6
Certo Liquid (2 pouches)	2.5	4	7.5	8
Freezer Jelly:				
Certo Liquid	1.5	2.5	5	5
BLUEBERRY				
MCP	2.25	4.5	6	7
BLACKBERRY				
Slim Set	3	6	4.5	8
Certo Liquid (2 pouches)	2	4	7.5	8
MCP	1.5	3	4.5	5
BLACK CURRANT				
Certo Regular	2	3	4	5
Certo Liquid	3	5	7	8
RED CURRANT				
Certo Regular	3.75	6.5	7	9.75
Certo Liquid	3	5	7	8

152

SUGARLESS JELLY

Manufacturer (1 batch=1pkg/bottle)	Fruit lbs.	Prepared (cups)	Sugar (cups)	Yield (cups)
Slim Set				
Raspberry	2.5	4	sweetener	4
Blackberry	2	4	sweetener	4

NUTRITION

per 3/4 c or 100 g	Potassium (mg)	Iron (mg)	A (mg)	B1 (mg)	B2 (mg)	C (mg)
Straw	140	0.9	8	0.03	0.05	62
Rasp	170	1	7	0.02	0.06	25
Blue	65	0.7	20	0.02	0.02	21
Blk Cur	300	1.3	23	0.05	0.04	170
Red Cur	230	0.9	6	0.04	0.03	35
Goose	200	0.6	34	0.02	0.02	34
Blackberry	190	0.9	45	0.03	0.05	17
Apple	130	0.4	10	0.03	0.03	11
Orange	183	0.1	21	0.07	0.02	53
Cranberry	71	0.5	5	0.03	0.02	11
Grapefruit	129	0.1	26	0.04	0.02	38
Grape	187	0.2	7	0.05	0.03	11

Courtesy Lucerne Foods

ACKNOWLEDGMENTS

I would like to sincerely thank my sister-in-law, *Rhonda Driediger* for all her enthusiastic and cheerful help. Without her, this book would have been on hold another year!

SPECIAL THANKS...

To my family for their patience.

To *June Driediger* for writing her story about the farm. It may be a short piece, but it took a lot of reminiscing.

To *Darlene Grant* of Bernardin of Canada, Ltd. for graciously supplying me with valuable information and recipes.

To *Debbie Hendersen* of E. S. Cropconsult, Ltd. for her insight as to the future of pesticide use.

To *Wanita Murphy* of Murphy Christmas Tree Farms for supplying me with research material and many great recipes.

To *Lorraine Jewell,* my neighbor, for cheerfully assisting me with writing this book.

To *Lynda Brown,* a Creative Memories representative who graciously coordinated all the pictures.

RECIPES SUBMITTED BY:

Customers of Driediger Bros. Farms Ltd.
Subscribers of all the local newspapers in the Lower Mainland
B.C. Raspberry Growers Assoc.
South Alder Farms Ltd.
North American Blueberry Council
Bernardin of Canada, Ltd.

INDEX

Blackberries

Blackberry and Apple Jelly 129
Blackberry Betty 77
Blackberry Crunch 74
Blackberry Jam 124
Blackberry Pie 102
Blackberry Roly-Poly 121
Kentucky Jam Cake 60
Spiced Jellied Blackberries 120

Blueberries

Blue Monday Cake 62
Blueberries Romanoff 85
Blueberry Betty 78
Blueberry Buckle Cake 76
Blueberry Cheesetart 104
Blueberry Clafoutis 106
Blueberry Cobbler 76
Blueberry Coffee Cake 62
Blueberry Cointreau 119
Blueberry Cottage-Cheese Cake 63
Blueberry Grunt 73
Blueberry Marmalade 132
Blueberry Milkshake 141
Blueberry Muffins 69
Blueberry-Orange Bread 137
Blueberry Pie One 91
Blueberry Pie Two 93
Blueberry Relish & Syrup 58
Blueberry Rhubarb Buckle 75
Blueberry Salsa 53
Blueberry Sauce 87

Cherry and Blueberry Pie 90
Foamy Blueberry Sauce 88
Fresh Blueberry and Lime Mold 47
Fresh Blueberry Cake 63
Fresh Blueberry Flan 108
Frozen Blueberry Parfait 84
Gingered Fresh Blueberry Compote 85
Grandma's Scones 69
Homemade Blueberry Pancakes 140
Homemade Blueberry Waffles 139
Lattice-Topped Blueberry Pie 101
Lemon Sauce for Fresh Blueberry Cake 89
Microwave Blueberry Pie 92
Mrs. Flaherty's Blueberry Loaves 137
Norwegian Blueberry Soup 50
Old Fashioned Blueberry Muffins 70
Peach Blueberry Pie 94
Spiced Blueberry Jam 128
Steamed Blueberry Pudding 82
Yogurt Blueberry Muffins 70

Currants, Black & Red

Apricot Red Currant Jam 123
Berry Medley Soup 49
Black Currant Liqueur 142
Cassis Sauce 55
Chilled Currant Soup 51
Currant Pie 101
Fresh Currant Ice 142
Fresh Currant Jelly 130
Fresh Glazed Currant Tarts 103
Old Fashioned Red Currant Jelly 130
Red Currant and Orange Jam 125
Red Currant Glaze 55

Gooseberries

Canadian Spiced Gooseberries 57
Frozen Gooseberry Fool 85

Gooseberry Charlotte 77
Gooseberry Chutney 56
Gooseberry Crumble 73
Gooseberry Curd 57
Gooseberry Dumplings 80
Gooseberry Fool 86
Gooseberry Jelly 131
Gooseberry Pie 92
Gooseberry Sauce 54
Green Gooseberry Jam 124
Mom's Gooseberry Jam 127
Stewed Gooseberries 111

Mixed Berries
Amaretto Cream 43
Anna's Berry Topping 89
Anna's Festive Summer Fresh Cake 59
Berry Bounty Bowl 114
Berry Cheesecake 65
Berry Cobbler 74
Berry Medley 84
Berry Pudding 81
Berry Shake 141
Berry Shortcakes 67
Blaspberry Pie 99
Blaspberry Squares 113
Blubarb and Raspberry Jelly 129
Deep-Dish Mini-Pies 93
Dessert Pizza with Fresh Fruit 110
Double-Crunch Bumbleberry Crisp 72
Frozen Berry Delight 114
Frozen Berry Dessert 118
Frozen Berry Mousse 117
Fruit Medley Clafoutis 107
Light Peach & Berry Crisp 71
Lorraine's Berry Dip 43
Low-Cal Berry Trifle 79

Natural Summer Fruit Jam - Chart 122
Old-Fashioned Cheesecake 66
Orange Poppy Seed Dip 43
Poppy Seed Dressing 46
Raspberry and Red Currant Jelly 131
Spirited Fruit 133
Strawberry and Red Currant Jam 126
Summer Pudding 81
Triple Berry Crisp Pie 99
Very Berry Spreadable Fruit 134

Raspberries
Best Frozen Raspberry Jam 125
Fresh Raspberry Salsa 54
Fresh Raspberry Vinaigrette 45
Low-Cal Raspberry Spread 88
Mile High Raspberry Pie 94
Old Fashioned Raspberry Bread Pudding 80
Raspberry Baked Apples 119
Raspberry Bombe 120
Raspberry Chicken 52
Raspberry Cream Pie 102
Raspberry Crisp in Microwave 71
Raspberry Delight 112
Raspberry Jam 125
Raspberry Jammer Bars 113
Raspberry Peach Pie 91
Raspberry Picnic Bread 138
Raspberry Pudding 83
Raspberry Sauce 88
Raspberry Supreme Dessert 116
Raspberry Swirl Torte 108
Raspberry Tarts 103
Raspberry Vinegar 136
Raspberry Wine Vinegar 136
Red Raspberry Vinaigrette 56
Sugarless Raspberry Jam 128
Very Raspberry Pie 100

Strawberries

Apple Strawberry Jelly 132
Avocado Ring with Strawberries 47
Cream Filling (for Strawberry Cream Pie) 98
Creamy Strawberry Blintzes (with Crepes) 115
Diabetic Strawberry Spread 134
Fresh Strawberry Pie 98
Fried Strawberries 111
Frozen Strawberry Cream 117
Golden Punch 143
Iced Strawberry Soup 49
Italian Strawberry Ice 142
Lattice Strawberry Tarts 104
Microwave Strawberry Jam 127
Our Family's Favorite Strawberry Shortcake 68
Spreadable Strawberries 133
Strawbarb Pie 100
Strawberries and Cheese 140
Strawberries and Cream Cake 61
Strawberry (Butter) Dressing 44
Strawberry (Spinach) Dressing 45
Strawberry & Avocado Salad 46
Strawberry & Rhubarb Clafoutis 107
Strawberry and Tofu Fool 86
Strawberry Barbeque Sauce 56
Strawberry Betty 78
Strawberry Bread 139
Strawberry Buttermilk Soup 50
Strawberry Cheesecake 64
Strawberry Chocolate Dip 44
Strawberry Chocolate Pie 95
Strawberry Cream Pie 97
Strawberry Dream Salad 48
Strawberry Dressing 45
Strawberry Glace Pie 96
Strawberry Hard Sauce 87
Strawberry Meringue Torte 109

Strawberry Mint Salsa 53
Strawberry Party Punch 143
Strawberry Rhubarb Conserve 135
Strawberry Rice Pudding 82
Strawberry Salad Supreme 48
Strawberry Snowy Pie 97
Strawberry Soup 111
Strawberry Supreme Pie 95
Strawberry Velvet Sauce 87
Strawberry Vinegar 135
Tropical Jam 126

"Happiness grows at our own firesides and is not to be picked in stranger's gardens."